A SHORT HISTORY OF
SHEEP
IN NEW ZEALAND

Richard Wolfe

RANDOM HOUSE
NEW ZEALAND

National Library of New Zealand
Cataloguing-in-Publication Data

Wolfe, Richard.
A short history of sheep in New Zealand / Richard Wolfe.
Includes bibliographical references and index.
ISBN 10: 1-86941-820-4
ISBN 13: 978-1-86941-820-5
1 Sheep—New Zealand—History. 2. Meat industry and
trade—New Zealand—History. I. Title.
636.300993—dc 22

A RANDOM HOUSE BOOK
published by
Random House New Zealand
18 Poland Road, Glenfield, Auckland
www.randomhouse.co.nz

First published 2006

©2006 text, Richard Wolfe; images as credited on page 219

The moral rights of the author have been asserted

ISBN-13: 978 1 86941 820 5
ISBN-10: 1 86941 820 4

Cover and text design: Trevor Newman
Cover image: Meat & Wool New Zealand and the New Zealand
Sheepbreeders Association

Printed in China by Everbest Printing Co Ltd

CONTENTS

INTRODUCTION 8

I
THE ORIGIN OF SHEEP 12

II
WHAT IS A SHEEP? 24

III
SHEEP ARRIVE IN NEW ZEALAND 30

IV
OUR FIRST FLOCKS 44

V
SHEEP BREEDS 74

VI
PRESERVING AND EXPORTING MEAT 96

VII
HOW TO LOOK AFTER SHEEP 122

VIII
TWENTIETH CENTURY SHEEP 142

IX
SHEARING 170

X
SHEEP FUTURES 188

GLOSSARY 195

BIBLIOGRAPHY 213

INDEX 220

INTRODUCTION

There aren't nearly as many sheep in New Zealand as there once were, but they still outnumber the human population by about ten to one. The only animal of any size inhabiting the country in greater numbers is an unwelcome import, the Australian possum, a pest that ravages what remains of our native forests. When sheep were first introduced here, they had a willing ally in the early European settlers, who enthusiastically dealt to extensive areas of forest on their behalf with axe and match before sowing grass seed in the ashes.

Although it took some time for flocks to thrive in New Zealand, now it might be described as a country ideally suited for sheep. For the first half century or so after their introduction, it was the sheep's wool that was in demand, but the arrival of refrigerated shipping meant New Zealand gained another valuable export in frozen meat. As a result, the beginning of the twentieth century saw both farm and sheep numbers grow steadily, with the animal's population peaking in 1982 at 70.3 million.

Despite their relatively short history in New

Zealand, sheep were one of the first animals to be domesticated. Some countries haven't taken to the animal, and vice versa, but their woolly forms are hard to avoid here. New Zealanders tend to take sheep for granted, and have been amused to discover that their most ubiquitous ungulate has also become something of a tourist attraction.

Not all sheep are the same. Thanks to human intervention, theirs is now a diverse species whose members have developed particular and distinctive qualities. If sheep farming was once mostly hard work, it is now equally a science, demanding the selection of breeds which best suit the conditions and the market.

Sheep have long been a mainstay of this nation's economy; indeed there was a time when they *were* the economy.

Sheep may no longer dominate New Zealand farms like they used to — for today's agrarian sector is nothing if not diversified — but meat and wool remain major players in the country's economy, and there will always be a place for their producers in our paddocks. They have enriched the language as well as the table, and been the butt of many jokes. They also represent a link with our rural roots in an increasingly urbanised society. It's hard to imagine New Zealand without them.

These sheep were no doubt more interested in getting to the greener grass on the other side than admiring this remarkable (and unidentified) tribute to the bridge builder's art.

Above: A modern shepherd and his flock, perhaps somewhere near the home of the first domesticated sheep, in Eurasia.

I

THE ORIGIN OF SHEEP

Sheep as we know them first appeared on earth about 2.5 million years ago, in Eurasia. While the modern animal now leaves well-worn and well-manured paths across the paddocks and hillsides of New Zealand, the evolutionary journey taken by its ancestors is not so obvious. It began in mountainous regions where fossil remains were destroyed by erosion and other natural forces. But what is known is that those early sheep were giants, as big as oxen.

By the end of the most recent Ice Age, about 10,000 years ago, sheep had down-sized considerably. The next big step — at least as far as humans were concerned — was domestication, a process that began at the end of the Mesolithic, or Middle Stone Age, between 12,000 and 15,000 years ago. But sheep have never been noted for innovation, and by coming under the influence of humans they were following a path already prepared by two other animals.

The first animal to be domesticated was man's best friend, the dog, which descended from the wolf. The latter animal already had much in common with man, for both had evolved as social hunters, preying on the same herds of large mammals. About 12,000 years ago this association was strengthened, perhaps initially by the taming of young animals that came into contact with human hunters. Those pups were likely to have had cuter and more endearing features than feral wolves, and several generations of breeding and rearing away from the wild species saw the evolution of a distinct type of tamer animal — the dog. It formed a mutually beneficial partnership with humans; one gained a home and a regular source of food, while the other now had a companion and an assistant. That assistant would later help with the control and taming of other species.

THE ORIGIN OF SHEEP ♈ 15

Above: The wolf, Canis lupis, *was ancestor of the dog, which later assisted man in the control of sheep. But the farmers' need to protect their flocks led to the elimination of large numbers of the world's wolves.*

> ## Woolly words
> WOLF IN SHEEP'S CLOTHING: An enemy disguised as a friend. Jesus warned (Matthew 7:15): 'Beware of false prophets, who come to you in sheep's clothing but inwardly are ravenous wolves.' A consequence was alluded to in Acts 20:29: 'For I know this, that after my departing shall grievous wolves enter in among you, not sparing the flock.' The same theme was also the subject of one of Aesop's fables, in which a wolf disguised himself in the skin of a sheep so that he would be shut up in the fold with the flock at night. The trouble was that the shepherd returned later to get some meat for his table, and killed what he assumed was a sheep — only to discover it was the not so cunningly disguised wolf.

The second animal to be domesticated was probably the goat. One animal authority, the late Roger Caras, claims that this took place during the Middle Stone Age, and that it helped shape the history of man 'as perhaps no other single event or accomplishment ever has'. Early hunters had followed herds of both goats

THE ORIGIN OF SHEEP ♈ 17

Above: The goat, Capra aegagrus hircus, *was probably the second animal to be domesticated — after the dog and prior to the sheep.*

and sheep as a regular source of meat, and thus began the process of human control. At first, goats were likely to have been of more use because of their wide-ranging browsing abilities. Their willingness to munch bitter, scrubby and thorny vegetation that was not palatable to other species, such as sheep, assisted early farmers in the clearance of land after forests had been cut down and burnt. The goat's ability to adapt to harsh conditions and thrive at higher altitudes than other animals also enabled man to extend his own range of habitation into desert and mountainous regions.

The domestication of sheep followed soon after that of the goat. The wild ancestor of the modern sheep is believed to have been the mouflon, whose habitat extended from Asia Minor to central Europe. An important part of this domestication was man's own development from hunter to herder, assisted by the dog which responded to instincts inherited from wolves and 'herded' prey to single out the weak, old or otherwise vulnerable. The obedient dog was now rewarded with food by its grateful owners, and so had increasingly less need to respond to its ancient feral urges. Sheep were thus drawn into a confined human environment where controlled breeding could take place. As with dogs, a domestic race of sheep — distinct from the original wild stock — began to emerge, and the earliest evidence for this dates from about 9000 BC in northeastern Iraq.

The earliest domestic sheep was similar to its wild ancestors, but smaller. One of the most significant differences was the horn, which had evolved originally for fighting purposes, both as a weapon and shield, and may have also helped the animal to balance on rocky terrain. Domestication of sheep led to changes in the shape of the horn, and also to its disappearance in ewes — probably as a result of efforts to produce an animal that was easier to manage. At the same time, tails became fatter and longer, while the coat changed both in woolliness and colour. The outer coat of the wild sheep was stiff and hairy, and covered a short woolly undercoat. With domestication, the bristly outer hairs disappeared, and the fleece changed to consist entirely of the undercoat. The first sheep as we know them now were white, and were followed by mixtures with black and brown faces and legs, which were unlike any of their original wild ancestors. Sheep, like goats before them, were now able to transform grass and other vegetation into food and resources for the benefit of man. In time they were selectively bred for certain characteristics, as providers of milk, wool and meat, which resulted in the appearance of distinct breeds by 3000 BC. One of those products, milk, would also prove to be handy for man's taming of other potentially useful species.

Flock of mankind

From its centre of domestication in Iran and Iraq, sheep spread out across the globe; herded westwards into Europe, and north and eastwards into Asia. In view of the timing and location of the animal's domestication, it is hardly surprising that the *Bible* features 247 references to sheep and the people who tended them.

Shepherding is the second form of employment mentioned in the Old Testament, although the first person to take it up was also named as the victim of the first murderer. Adam and Eve's son Cain was a tiller of the soil, while his younger brother Abel looked after sheep. When the pair made sacrifices of their respective produce, God was less impressed with Cain's offering of grain. The jealous Cain then killed his brother, and when God asked after Abel's whereabouts, Cain replied, 'I know not: Am I my brother's keeper?'

As well as providing such earthly needs as meat and wool (which may have been the raw material for Joseph's coat of many colours), and serving sacrificial duties, sheep acquired metaphorical status in the *Bible*. Mankind itself was referred to as a flock tended by the

THE ORIGIN OF SHEEP 21

Above: In a scene which may have changed little since Biblical times, Middle Eastern shepherds tend a mixed herd of sheep and goats beside an ancient stone wall in Jerusalem.

ultimate good shepherd, the Christian God. Thus, when Christ dispatched his disciples to the 'lost sheep' of Israel, they went as 'sheep in the midst of wolves'; and to counter the godless ways of men they were also required to be as wise as serpents and as harmless as doves. As for the sanctity of the Sabbath, which was given over solely to worship or rest, if a sheep had fallen into a pit on that day, it was permissable to rescue it. And on Judgement Day, all the nations would be separated as a shepherd divided his flock, with the sheep honoured by being on the right-hand side of God, while the goat was relegated to the left.

Sheep enjoyed an exulted status in the *Bible*, at least compared with goats, but both animals were recognised by the ancients. A cluster of distant stars which resembled the head of a ram, complete with a spiral horn, became known as Aries, after the Latin word for the male sheep. This constellation is the first sign of the zodiac, heralding the astrological new year when the sun begins its cycle on the vernal equinox, about 21 March.

Later, the tenth sign of Capricorn was represented either by the figure of a goat, or the fore part of a goat combined with the hind part of a fish. The fore part of the word Capricorn is from the Latin for goat, which is also the root for the word 'caper', meaning a leap or spring, much as that animal does in the wild. By comparison, the word 'sheep' enjoys no such energetic or frolicsome connotations, with one late nineteenth century dictionary defining it as 'a weak, bashful, silly fellow'.

Back on earth and domesticated, goats and sheep would soon extend their range and serve an increasing number of cultures, eventually finding their way even to New Zealand.

> ### Woolly words
> SHEPHERD'S PIE: Traditional dish with a meat base and a mashed potato topping. Of less concern to sheep is a version made with beef and more properly known as a cottage pie.

Above: As its name suggests, the most distinctive feature of the bighorn mountain sheep, Ovis canadensis, *is the massive pair of horns which spiral backwards from the top of its head. If this specimen appears aloof, it may be because the species had nothing to do with the development of the domestic sheep.*

II

WHAT IS A SHEEP?

From the outset the sheep was an extremely successful animal, with the high insulation qualities of its wool enabling it to adjust to climates ranging from hot deserts to arctic conditions. Later, use of this wool also enabled humans to migrate to less hospitable regions. And down at ground level, the sheep's mobility was increased by the versatility of its cloven hooves, which allowed it to take to widely different types of terrain.

Being warm-blooded vertebrates that suckle their young, sheep belong to the class of animals known as mammals. Further, their cloven hooves qualify them as even-toed ungulates and members of another order (Artiodactyla), along with pigs and camels. But unlike them, the sheep chews cud, or ruminates, and so belongs to a sub-order (Pecora). There are three families of these ruminants: giraffes, deer and the cattle family (Bovidae), and sheep belong to the third, together with goats, cattle and antelopes. The cattle family is distinguished by hollow horns that grow continuously, unlike the antlers of deer which are solid and are shed and regrown annually.

Unlike other members of this group the sheep has a cleft upper lip. This allows it to graze more closely and selectively than cattle, so the two animals can happily co-exist rather than compete for food in the same immediate environment. In some cases, goats are distinguished from sheep by beards, while there are differences in their horns and tails, and the males have odoriferous glands beneath their tails. Domesticated goats are also much more likely than sheep to undo the efforts of man and revert to a feral state, when given the chance.

Being a ruminant, a sheep has a stomach consisting of four compartments. Food begins its journey in the largest of these, the rumen or paunch, and then passes through to the adjacent reticulum, which has a honeycombed lining that is more familiar to some humans as tripe. Semi-digested food is periodically returned to the mouth to be chewed again as 'cud', a digestive process which requires regular regurgitation and belching, as well as special dental equipment. In their lower jaw sheep have eight front teeth — canines and incisors — for cutting purposes, which bite against a horny pad, and there are no front teeth in the upper jaw. The food then proceeds to the back teeth which have special ridges for grinding. Jaws move widely from side to side, and because the lower jaw is narrower than the upper, the sheep chews on only one side at a time.

Natural selection

Today's domestic animal is one of eight species of sheep, *Ovis aries*. Of the seven surviving wild members of the genus Ovis, the mouflon (*O. orientalis*) is probably the ancestor of our domestic animal. Perhaps the best known feral member is *O. canadensis*, the American bighorn or mountain sheep, which does not feature in the ancestry of domestic sheep.

Today there are at least 326 fully recognised breeds of sheep, although there are many more varieties which are mostly geographic alterations. The countries with the greatest number of their own sheep breeds are the former USSR, which developed 58 breeds to suit its vast range of climates and cultures, while France and Italy claim 47 and Wales, Scotland and England 41. New Zealand meanwhile has managed to do well with some two dozen imported breeds. From these has developed a relatively small number of distinctive breeds, including the New Zealand Romney, Coopworth, Perendale, Corriedale, New Zealand Halfbred, Drysdale, Borderdale, South Suffolk and South Hampshire.

Wild sheep have survived in various parts of the world, and are distinguished from their domesticated relatives by being more slender and having short tails and large horns. They have black, brown, grey and white areas in the coat, which no doubt provide camouflage in their natural habitat, while the coat consists of a hairy outer layer over a woolly undercoat.

Human intervention resulted in at least one other significant development, and one which is probably lost on the sheep. The skull of the modern animal indicates a decreased brain capacity and a smaller diameter eye socket when compared to that of its wild ancestor. While domestication of animals was intended initially to make things easier for humans, it seems that life

Above: The result of some 15,000 years of human intervention: a flock of New Zealand sheep.

Woolly words
WOOLLY THINKING: Lacking intellectual rigour. Also woolly-headed thinking, which is not the same thing as fuzzy logic.

away from the wild had a similar effect on sheep. Their relatively worry-free existence, with all concerns taken care of by kindly farmers, has reduced their need for the sensory faculties required for survival in the wild.

Above: A busy pastoral scene at the Waimate North mission station, established in 1830. By then sheep and other farm animals brought to New Zealand by missionaries were well established around the Bay of Islands.

III
SHEEP ARRIVE IN NEW ZEALAND

Humans became increasingly dependent on domesticated sheep, and took them with them when establishing new settlements, as in the Americas following Columbus's discoveries in 1492. By the beginning of the eighteenth century, huge flocks of sheep were grazing in the British Isles and Europe where they provided a vital source of food and clothing for a growing population. With the invention of the automated loom and the building of mills for the spinning and weaving of wool and cotton, an agrarian economy began to be transformed into an industrial one.

Sheep were introduced to New Zealand in 1769, arriving with Lieutenant James Cook on his ship *Endeavour*. They were not allowed to venture ashore, which was probably just as well, but they did have the distinction of being the first of their breed to enjoy the primitive pastures of this country. Crew members were landed to cut fresh grass on at least two occasions, at the Bay of Islands and later in Queen Charlotte Sound. Four years later, sheep finally came ashore, but not for long. On 23 May 1773, during his second voyage, on the *Resolution*, Cook landed in New Zealand with a ram and a ewe, the only survivors of a flock of six taken aboard at the Cape of Good Hope.

He liberated the pair in the hope that they would be of benefit to local Maori, but was disappointed at the failure of his attempt at acclimatisation: 'Last Night the Ewe and Ram I had with so much care and trouble brought to this place, died, we did suppose that they were poisoned by eating of some poisonous plant, thus all my fine hopes of stocking this Country with a breed of Sheep were blasted in a moment.' Those unfortunate sheep had endured a voyage through the Antarctic Ocean and been nursed through scurvy in Dusky Bay, only to be killed by the poisonous berries of the tutu plant. They were the first, but hardly the last, sheep to die this way in New Zealand. During his

Above: *A 1783 engraving of the portrait of Captain James Cook painted by Nathaniel Dance in 1776, the year Cook began his third voyage to New Zealand. On all his visits to this country Cook brought sheep, but none survived.*

earlier *Endeavour* voyage, Cook had introduced pigs to this country, and unlike sheep they took to the fern and multiplied.

On his third voyage, on the *Resolution* and *Discovery*, Cook again took sheep aboard. They were part of an extensive menagerie, and at Ship Cove on 12 February 1777 surgeon's mate David Samwell described his vessel as a 'second Noahs Ark' when its cargo of horses, cattle, sheep, goats, peacocks, turkeys, geese and ducks went ashore for feeding, 'to the great astonishment of the New Zealanders'. The sight reminded him of rural England, a nostalgic reflection that may have been assisted by the spruce beer that the men brewed here. The sheep on this voyage had also been picked up at the Cape of Good Hope, but had grown weak by the time they reached New Zealand and died soon after being put ashore. It would be nearly another four decades before any members of the species *Ovis aries* would survive their introduction to the country.

Sheep were in fact well beaten to New Zealand by another early domesticated animal, the goat. That hardy animal had proved easy to keep alive on ships, and been introduced to the West Indies and islands in the Pacific and Atlantic by Spanish and Portuguese navigators from the mid-fifteenth century. When the *Endeavour* sailed

> ### Woolly words
> When the Second Witch in Shakespeare's Macbeth uttered 'Paddock calls' she did not mean an enclosure for sheep but an alternative name for a toad. However, sheep were referred to in another of Shakepeare's plays, The Two Gentlemen of Verona, which included a shepherd, sheep's horns, mutton and even the word 'baa'. The bard also demonstrated his knowledge of the behaviour of that animal with the lines 'A silly answer and fitting well a sheep' and 'Indeed, a sheep doth very often stray'.

from Plymouth on 16 August 1768, it also took a goat to supply fresh milk for the officers. On Cook's second voyage two animals — a male and female — were put ashore on the western side of Arapawa Island, in Queen Charlotte Sound, in the hope they would survive in the local bush. Perhaps they did, for there are feral goats on that island today, and some may be descended from the pair liberated by Cook.

Following the publication in Europe of Cook's voyages, the British Government decided to establish

Above: It is possible that two hardy goats put ashore at Arapawa Island in the Marlborough Sounds during Cook's second voyage may have survived, and are the ancestors of today's feral population.

a penal colony in New South Wales. The First Fleet arrived at Botany Bay, south of Sydney, in January 1788 and, along with convicts and other settlers, it landed about 100 sheep. The animals had been picked up in South Africa and were of the native Cape variety, with extremely large tails that provided fat which local farmers used as a substitute for butter.

Merino magic

Six months later, only 29 from that first flock had survived their arrival in Australia, and more animals were imported from India. But none of these early types would be the basis for the large Australian sheep industry that followed; that honour would go to the Merino.

Originally from Spain, the highly sought-after Merino had been jealously protected in its homeland, but from the early eighteenth century it was introduced to other European countries, often as a personal gift of the King. From there its range was extended, eventually reaching Australia and New Zealand.

In 1789 Captain Henry Waterhouse, who had gone to New South Wales with the First Fleet, was sent to the Cape of Good Hope to obtain sheep. He was a reluctant stock purchaser, regarding that duty as somewhat beneath his rank as an officer, but he did acquire 32 Merinos. Twenty-nine of them survived the trip back to Australia, where they formed the basis for other flocks. Captain John McArthur, for example, obtained three rams and five ewes from Waterhouse's stock. He was so impressed by how well they took to their new home that he went to England and obtained further specimens, which had been given to George III by the King of Spain. McArthur was later involved with the Rum Rebellion in Sydney, when British soldiers mutinied against New South Wales Governor William Bligh, of *Bounty* fame, and was banished to England. He returned south in 1816 and made a fortune from sheep, and is now regarded as the founder of the Australian wool industry.

Another who saw a great future for sheep and wool growing in Australia was the Reverend Samuel Marsden. Born in Yorkshire, the son of a blacksmith and farmer, he was appointed second chaplain in New South Wales, arriving at Port Jackson, Sydney, in March 1794. He took up farming near Parramatta and in 1797 he also obtained sheep from Waterhouse, and within seven years was reported to have 1200 such animals of 'a high class'. He mated his Merino ram with hairy ewes to improve the fleece, and when he went to Britain in 1807 he took 165 lb (75 kg) of his wool with him.

It was enthusiastically received by Yorkshire manufacturers who had been unable to obtain their usual supply as a result of the recent French invasion of Spain. Marsden's wool was made into 40 yards (36.6 m) of worsted cloth, and from this he had a suit tailored for himself. He wore it when visiting George III, who was so impressed he requested a coat for himself made from the same material. In return

Above: Originally from Spain, the Merino was introduced to New Zealand from Australia. It remains the most numerous sheep breed in the world, and dominated this country's flocks until the end of the 19th century.

'Farmer' George gave Marsden five Merino ewes in lamb from his flock at Kew, to help promote a wool-growing industry in New South Wales.

Four ewes and two lambs from the Royal flock survived the voyage to Australia. Later, eight Merinos, along with other animals, travelled with Marsden across to New Zealand, arriving at the missionary settlement of Rangihoua in the Bay of Islands on 23 December 1814. These sheep were soon supplemented by twelve more, purchased by missionary William Hall. He moved them to better grazing land across the bay near Waitangi, but in January 1817, after a quarrel among local Maori, he had to abandon his sheep and take his family to the security of Rangihoua. He later located a small number of survivors of that flock and, in relieving them of their wool, probably qualified as this country's first shearer.

In 1819 the missionaries established a second settlement, at Kerikeri, west of Rangihoua. Major Richard Cruise, in charge of troops on the naval vessel *Dromedary*, visited the small community the following year and observed its flock of sheep. He saw no natural grass but assumed there was a 'nutritive herbage' among the fern because the animals appeared to be in good condition. By at least 1824 this flock was being shorn, although missionary John King reportedly had to do so with a pair of scissors. Even so he managed to produce more wool than the community needed for its own use and eleven bags were exported to Sydney, where the wool realised two shillings and six pence per pound.

Samuel Marsden regularly sent animals from

his farm base at Parramatta to the missionary establishments across the Tasman. Not long before he died in 1838, this son of a farmer made a perceptive comment on the future of New Zealand, informing a fellow missionary that sheep, cattle and horses would be 'the making of the country'.

> ## Woolly words
> LIVING ON THE SHEEP'S BACK: Strictly speaking, Australia — the world's largest wool exporter — has a much stronger claim than New Zealand to this condition. While this country depended on wool off the sheep's back (and elsewhere) for the latter half of the nineteenth century, the advent of refrigerated shipping led to meat becoming an even more valuable export. On the Sheep's Back was also the title of an exhibition celebrating the national woollen industry held at Te Papa Tongarewa, Museum of New Zealand, in 2006.

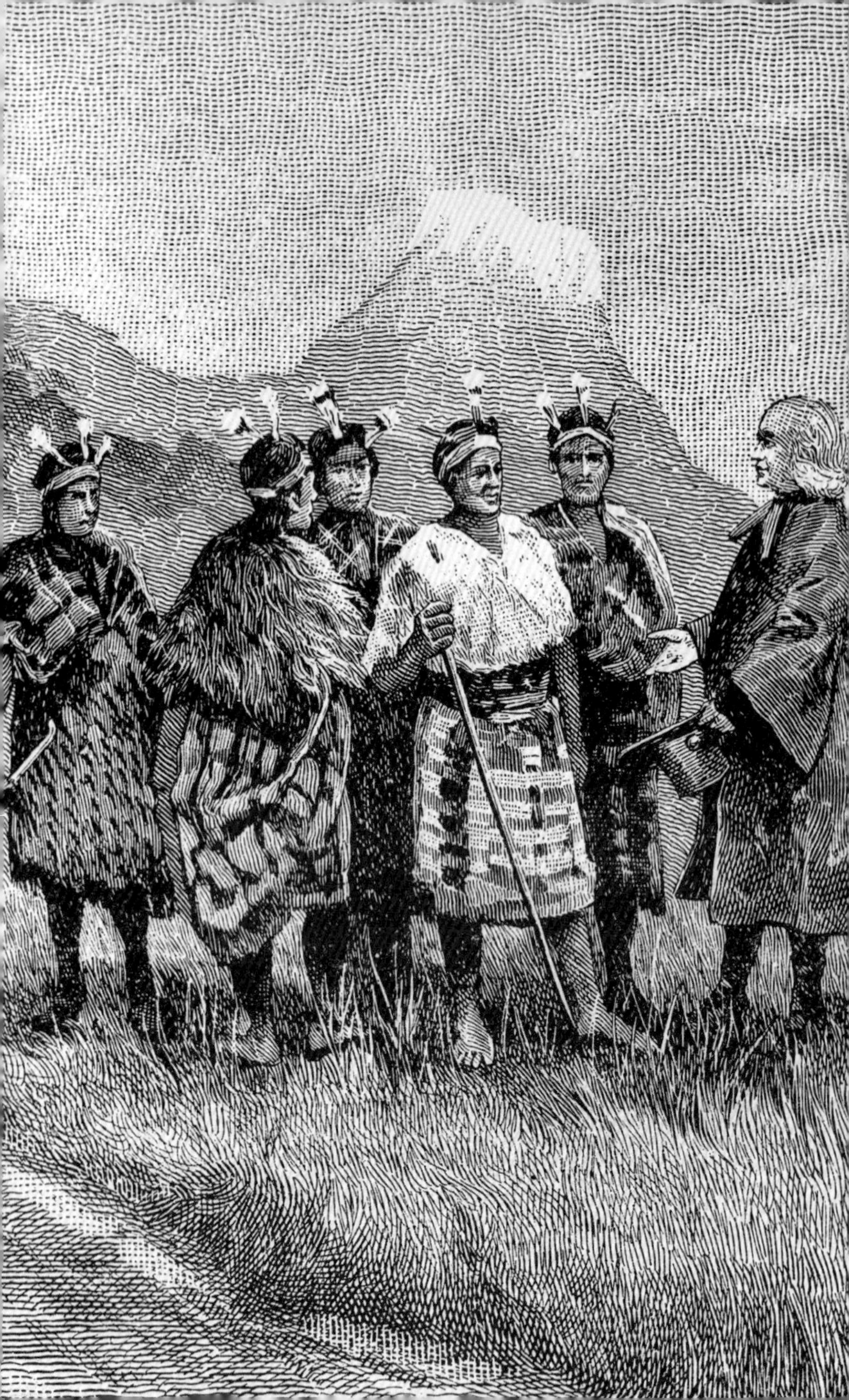

The Reverend Samuel Marsden and his entourage coming ashore at the Bay of Islands on 19 December 1814. On this, his first trip across the Tasman, Marsden not only introduced religion to the Maori, but also brought eight Merinos, along with other animals.

Above: Painting by Samuel Charles Brees in 1847 of land near Ngaio, in what was known as the Parerua (Porirua) Bush, cleared by pioneer agriculturalists Charles Clifford and his cousin William Vavasour. The pair had arrived in Wellington in 1842, and Clifford later took up sheep farming in the Wairarapa, Marlborough and North Canterbury.

IV

OUR FIRST FLOCKS

Mana Island, some seven kilometres off shore from the entrance to Porirua harbour, near Wellington, was once a popular spot for whaling ships, which called in to stock up on potatoes and pigs supplied by Maori agriculturalists on the mainland. In 1834 Scotsman John Bell Wright arrived at the island, bringing cattle and some 100 Merino sheep from Sydney.

Depending on what constitutes a flock, Wright's sheep may have been part of the first to be brought to this country. Their arrival might also justify Mana Island's claim as the birthplace of the New Zealand sheep industry, as the enterprising Wright was — like John King — an early wool exporter, sending a few bags to Sydney in 1835.

At Mana, whalers could now supplement their diet with mutton and lamb. When the first New Zealand Company settlers arrived in the Wellington area in 1840, the sheep they saw around Petone were probably descended from Wright's pioneering flock. They may have been firmly established here by now, but this country's suitability for such animals wasn't yet apparent to some.

In May 1840 an English publication anticipating the spread of the Anglo-Saxon race to the Southern Hemisphere suggested that Australia might become the sheep farm of Europe, while the islands of New Zealand could be 'the granary of Australia'. Both grain and sheep would soon prove to be of great importance to the New Zealand economy, and on 5 October 1840 there was something of a landmark event in connection with the latter industry. Heading the list of merchandise at an auction at Te Wahapu in the Bay of Islands were 20 sheep, which may have had the honour of being the first livestock to be sold by that means in the country.

Sheep were here to stay but, because the pioneers who ran the first flocks undertook a range of agricultural activities, they did not yet qualify as dedicated sheep farmers. That would all change in the early 1840s with the arrival of the pastoralists, and the realisation that sheep farming was probably the best — if not the only — means of utilising this country's large tracts of hilly and mountainous terrain. At that time sheep offered only one economically viable product for export — wool. Unlike meat at that stage, it had the great advantage of being non-perishable and thus was suitable for transporting on lengthy sea voyages.

Pastoral pioneers

The age of pastoralism arrived in 1843 when Charles Bidwill, originally from Exeter, England, brought some 1600 sheep over from Sydney to Nelson. He and three others — Charles Clifford, from Liverpool, his cousin William Vavasour, and the Hon. Charles Petre — had heard reports of the suitability of the Wairarapa for sheep, and so arranged leases on land. But access to this region in the southeastern corner of the North Island was difficult, as it was cut off

48 A SHORT HISTORY OF SHEEP

Above: Soon after his arrival in Wellington in 1844, Frederick Weld became involved in a sheep run in the southern Wairarapa, and in 1847 he began exploring farming possibilities in the South Island. A keen watercolourist, he painted Canterbury Plains, Waimakariri *in 1850.*

from Wellington and the west by the Tararua Range and its southern extension, the Rimutaka Range.

Bidwill set out for the land in April 1844 with 350 Merino sheep, driving them around the rugged coastal route from Port Nicholson (Wellington), past Baring Head. At times the animals had to be man-handled around the rocks, while the would-be graziers also had to negotiate an overflowing Lake Wairarapa, as well as swamps and areas of bush. Once they had arrived at their new home, the sheep had to deal with wild dogs and feral pigs, while the farmers endured the added discomfort of mosquitoes.

Another attracted initially to the prospects in the Wairarapa was Frederick Aloysius Weld, from Dorset. A cousin of Clifford and Vavasour, he arrived in Wellington in 1843. He quickly came to the conclusion that sheep farming was likely to be the most promising source of income in this country. He guessed correctly, and when profits from his pastoral activities enabled him to return to England, he published an influential pamphlet, *Hints to Intending Sheep-Farmers in New Zealand*, which eventually ran to three editions. In late November 1864, Weld became Premier of New Zealand, the sixth to that date, but he only held the post for a little less than eleven months.

In 1847 Clifford, Vavasour and Weld were attracted

by reports of open tussock country in the South Island. Leaving Wairarapa, they landed some 3000 sheep from Sydney at Port Underwood in Marlborough. Again they took a coastal route, and after a nineteen- day journey along Cloudy Bay to Cape Campbell, they occupied (or 'squatted' on) an area which later became Flaxbourne, near the township of Ward.

> ### Woolly words
> BEING PUT OUT TO PASTURE: When someone is taken off a job for reason of age. In colder climates where animals are kept indoors, they may be 'put out to pasture' for only part of the year. In New Zealand a pet sheep may be put out to pasture, under no (immediate) threat of being sent off to the works.

Road to wealth

Sheep were soon spreading throughout the country, south from Marlborough. Canterbury lacked the timber or geological resources of other regions but it had vast tussock grasslands suitable for sheep, and wool seemed the obvious road to wealth. Two local pioneers were the Deans brothers from Ayrshire, Scotland. William arrived at Port Nicholson on 21 January 1840 and, after investigating various parts of the country, he decided to settle on the Canterbury Plains.

His brother John arrived at Nelson on 25 October 1842, and the pair settled at a spot they called Riccarton, after their home parish back in Scotland. They named the local stream the Avon and established a farm with stock from Australia, including 61 cattle and 43 sheep, while two further trips were made in the late 1840s for another 1200 sheep.

At the beginning of the 1850s, the most populated province in terms of sheep was Nelson, with 92,014, while Canterbury was fourth with 28,416, trailing Wellington (64,009) and Otago (34,829). But within seven years Canterbury was leading the field, with 495,580 out of a national population of 1,523,324 — which also took into account eight animals on Stewart Island.

Two of Canterbury's most notable sheep farmers at this stage were lifelong friends from Devonshire, John Acland and Charles Tripp. They had both decided to give up legal careers in England and take up land in New Zealand, arriving in Lyttelton in October 1855. After gaining experience on established runs, they applied for land of their own, but had insufficient

Above: An unknown artist's record of the farm established in 1843 at Riccarton, Canterbury, by William and John Deans. In the foreground is the river they named the Avon.

funds to buy an existing station. Because all of Canterbury, including the foothills, was already taken they looked in the region of Mt Peel and the Rangitata River gorge, and established several stations in the area. The partnership lasted six years until the property was divided, with Acland retaining Mt Peel — which would later reach an extent of 100,000 acres (40,500 ha) — and Tripp taking Orari Gorge Station.

A hazardous life

As the South Island grew in economic importance, initially thanks to wool, a drought in Australia in 1850 encouraged sheep farmers to bring their money, stock and experience across to New Zealand. These graziers burned off areas of tussock and planted English pastures on the Canterbury Plains for their Merinos. This breed of sheep were good fossickers for food, and suited the South Island where they were less troubled by footrot than in the wetter North Island.

Footrot is a disease of the hoof which was treated with solutions of carbolic acid, caustic preparation or arsenic. Another serious problem for sheep was scab, which had come with Merinos from Australia and was in Nelson by 1845. Small burrowing insects caused eruptions on the sheep's skin which resulted in crusted

Above: An 1864 Australian illustration of shearing in an open-ended shed, a scene that was repeated on this side of the Tasman. A drover and his dog (left) rest while shearers and shed hands go about their work — all under the watchful eye of the runholder. The newly shorn sheep emerge on the right.

scabs, and hence the name of the disease. Irritation caused sheep to rub and damage the fleece and also lose condition, and the usual remedy was dipping with a mixture of either sulphur and tobacco, or lime. Scab peaked in the 1860s, and was officially declared eradicated in 1892.

There were plenty of other challenges for early runholders, and drastic ways of dealing with them. Blowfly maggots burrowed under the skin of the sheep, usually near the tail, while the gadfly (or bot or 'grub in the head') deposited eggs in the sheep's nostril. They could be dealt to variously with strong-smelling whale oil, tobacco wash or diluted carbolic acid. Then there was fluke (also called 'rot'), which infected the liver, and ticks, which pierced the skin of the sheep. Infestations of lice were cured by dipping after shearing, either in tobacco and sulphur, lime or a patent medicine.

Sheep could also be bothered by wild dogs, while the farmer needed to remove areas of the swordlike speargrass and the prickly and scrubby matagouri, or wild Irishmen, to make way for his flock in the tussock grasslands. He also needed to keep them away from tutu, which has been responsible for the greatest percentage of stock poisoned by plants in New Zealand — beginning in 1773.

> **Woolly words**
> THE GRASS IS ALWAYS GREENER:
> Dissatisfaction with the present situation. A sheep will frequently aspire to the grass on the other side of the fence, which may very well be greener.

Besides such hazards for sheep on the new pastures of New Zealand, just getting stock to the country was risky. In October 1848 the ship *Speed* arrived in Auckland from New South Wales with 100 sheep, but an equal number, including 40 fat cattle, died on the voyage. The situation had barely improved by the mid-1860s on three further shipments across the Tasman: the *Adelaide Bell* lost 120 sheep out of 180, the *H.L. Rutgers* lost 141 out of 171, and *The Western Star* lost 113 out of 151.

The wool kings

With the arrival of the pastoralists, the nation's sheep population increased rapidly, passing its first million in the mid-1850s. It then grew to 2.8 million in 1861, and reached 9.7 million in 1871, when Canterbury and

Otago each claimed nearly one third of New Zealand's total flock.

Although the price of wool fluctuated, in 1872 it became New Zealand's leading single export item in terms of earnings. It enabled the 'wool kings' to build themselves impressive homesteads and introduce the comforts of civilised living to their outback stations. Another settler who arrived at this time has been described by writer Gordon McLauchlan as this country's 'all-time best known sheep farmer', and was probably also the best-known author to have lived in New Zealand at the time. Samuel Butler was born in Nottingham and, after preparing for a career in the church, he left for this country with the intention of sheep farming.

He reached Canterbury in early 1860 to find, as had Acland and Tripp before him, that all the province's known sheep country was taken up. Exploring further afield, Butler found some 5000 acres (2000 hectares) of unclaimed tussock land between the rivers Rangitata and Rakaia, which became the nucleus of his station. He named it Mesopotamia, Greek for 'land between two rivers', after the ancient region in southwest Asia between the Tigris and the Euphrates and, coincidentally, near the birthplace of the modern sheep.

Butler built a homestead and lived here for three

The homestead and outbuildings at the heart of Mesopotamia, the pastoral run established in South Canterbury by Samuel Butler in 1860. This photograph was taken in 1871, seven years after Butler sold the property and returned to England, and the year before he published his novel Erewhon.

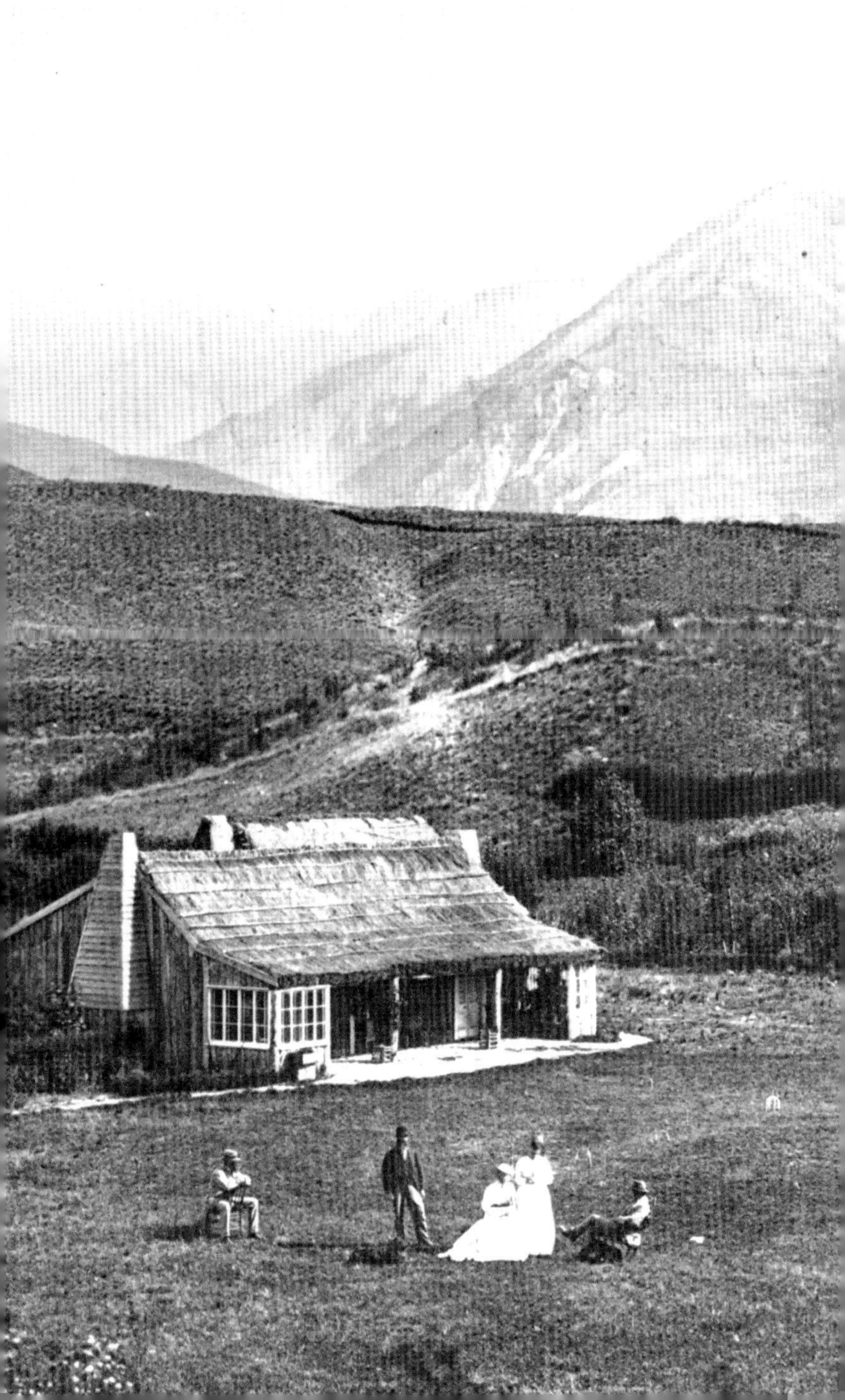

and a half years, enjoying the comfort of a piano, along with his books and paintings. In May 1864 he sold Mesopotamia for an impressive profit, and returned to London to devote himself to art — exhibiting at the Royal Academy — and writing, publishing his satirical *Erewhon* (an anagram of 'Nowhere') in 1872. He challenged the controversial views of Charles Darwin, as well as the conventional morality of his time, and must remain as the most remarkable person ever to take up sheep farming in New Zealand.

But when it came to sheep management, there were few as canny or as infamous as James Mackenzie. A native of Inverness, Scotland, he emigrated to Australia in about 1849 and then crossed to New Zealand, working his way from Nelson to Southland. In March 1855 a mob of about 1,000 sheep missing from the Levels Station in South Canterbury was traced westward to the huge high-country basin now known as 'the Mackenzie Country'.

Mackenzie was located with the lost flock and was caught by the overseer of the station, but he managed to escape. He reached Lyttelton, where he was arrested and charged with the theft of sheep, and on 12 April was sentenced to five years' imprisonment. He was pardoned after serving only nine months, during which time he escaped twice and was recaptured. Even while

Mackenzie was in custody, his highly intelligent dog, Friday, allegedly continued to drive sheep, without his master's instructions. In January 1856 Mackenzie left New Zealand for Australia. There is little documentary information about him — or his equally legendary dog — and it is almost impossible to separate fact from fiction. A bronze memorial to Friday's breed of working Collie dog was erected on the shores of Lake Tekapo by grateful farmers in 1968, while a statue of James Mackenzie has stood in the main street of Fairlie, South Canterbury, since 2003.

Woolly words

MUTTON DRESSED AS LAMB: Someone dressed in a manner more suited for a person much younger. Mutton was obviously less tasty than the meat of a younger lamb. Perhaps a reference to a butcher's tempting shop-window display. The term 'lamb-fashion' was also used in the nineteenth century.

Pasture pest

Wool was now a major export earner for the country, but in many regions the sheep's habitat was under threat. There was growing compensation from another imported animal, but it was one that rapidly became a major pest. Some sheep may even have had a chance to encounter this threat to its livelihood as early as December 1849, at the annual show of the Auckland Agricultural and Horticultural Society. The event was attended by Sir George and Lady Grey, and the livestock on show included Leicester sheep and a 'remarkably fine pair of rabbits'.

The latter animal had been deliberately introduced to this country from about 1838, but it was not until the late 1860s that rabbits became well established in Otago and Southland. From there they surged northwards, moving quickly through the cleared tussock land, while others swept southwards from Marlborough, all the time helping themselves to the sheep's food supply. As a result flock sizes at a number of southern stations plummeted, while the razing of vegetation on the light-soil pastures of the South Island hills would also cause further problems of erosion.

In 1876 the Government passed the Rabbit Nuisance Act, but it was ineffective. The official approach was

Above: Armed rabbiters at their camp in 1909. Low prices for sheep products in the late 19th century stimulated the commercialisation of rabbits in New Zealand, and the peak year for the export of their skins was 1894. In 1901 refrigeration enabled another six million rabbits to be 'removed' from the pastures of New Zealand, and exported as carcasses.

that the problem would be controlled by an industry, which would earn the country good money in the process. Rabbit skins were exported during the early 1870s and carcasses went later, thanks to the advent of refrigeration, but it seemed to have little effect on numbers. In fact, the rabbit population quickly outstripped that of sheep, if the export of about one million skins in 1877 and more than nine million in 1882 is any indication.

Rabbits may have appeared to be a valuable export, but the income gained was heavily outweighed by the economic waste caused by the displacement of sheep and the damage to vegetation and soils. It was not until major commercial activity surrounding rabbits was prohibited in 1952, when the export of rabbit skins and carcasses was banned, that the problem was taken seriously.

By 1961 the country had 161 rabbit boards dealing with the problem, but despite aerial poisoning, ground baits and shooting, the rabbit proved too prolific a breeder to be completely eradicated in New Zealand. Since the early 1950s, there have been several attempts to introduce the deadly disease myxomatosis, which is spread among rabbits by fleas. Research in Australia suggested the rabbit calicivirus disease (RCD) may be a more effective and humane means of control, and in August 1997 the disease was illegally introduced to New Zealand in an attempt to control this pest.

A Growing Clip

Sheep shearing was the economic highlight of the runholder's year. It was a summer job, and at first was carried out in the open. The sheep were washed in preparation, but a high death rate soon stopped that practice. Dedicated shearing sheds were built, some of them massive structures, and they became distinctive features of the rural landscape. At first sheep were shorn with blades, but machines reached this country from New South Wales by the end of the century, and were soon in general use.

The centres of world wool production were now shifting. Britain had earlier obtained most of its Merino wool from Spain, but that was gradually replaced by Germany, and then by sources much further afield. As a result, by 1890 two thirds of the wool imported into the United Kingdom was produced by the sheep of Australasia.

New Zealand's own production during this period was increasing rapidly, from 78.6 million lbs (35 million kg) in 1884 to 122.3 million lbs (55.5 million kg) in 1893, despite the problem with rabbits. This larger wool clip was mainly due to a growing sheep population (19.4 million by 1893), but there had also been a growth in the number of smaller flocks, and such farmers were better equipped than those with large runs to cope with the ravages of the rabbit.

While the bulk of New Zealand's wool was destined for Britain, a small amount — usually about two per cent of the total produced — was absorbed by industry in this country. The shipping of prepared flax fibre to Sydney had already qualified textiles as one of New Zealand's earliest exports, while locally grown flax was used to manufacture sacking bales to transport what was now the country's largest single export item — wool.

In the decade from 1884, wool was responsible for nearly half of the country's export receipts, although frozen meat, butter and cheese (made possible by refrigeration) now claimed an increasing share.

One of the major local users of New Zealand wool began business as the Canterbury Flax Spinning, Weaving and Fibre Company with the intention of processing flax fibre and turning it into sacks, scrim and woolpacks. But the cost of machinery forced a change of focus to the manufacture of woollen and flannel goods. The company's first blankets went on sale in 1876, and two years later the enterprise became the

DONALD & SONS, LIMITED

Sewing the Bale in No. 2 Press

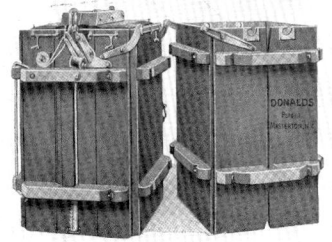

Showing Pressed Bale and Facility for Sewing.

We are sometimes asked how the operator manages to sew the bale along the back of the box in the No. 2 Press. The illustration above shows how this is satisfactorily accomplished. One hinge is made to conveniently uncouple, and the other is joggle eyed to enable the box to be swung around without being disengaged. This arrangement is part of our patent, and is a special feature used exclusively on our No. 2 Press. By its aid the sewing becomes a simple operation.

ENVELOPE BALES, fastened with metal clips or stitches, can be used with any of our Presses. We always like to know whether this method is to be adopted before we send out the Press, also what brand of bale grips is to be used; the reason for this is that with most bale grips a small " V " shaped piece has to be taken out of the door and back of the Press to enable them to be worked.

The above information relates to the No. 2 Press only; the No. 1 has special sewing facilities of its own.

STEEL LINERS
(PATENTED)

A recent addition to our Presses is that of adding Steel Liners in the journal boxes. These liners take the wear, and give a most excellent bearing between the journal and the box in which it works. This makes for saving in working effort and much longer life in the parts. It is our latest patent, and is looked on by ourselves, engineers, and users, as being an exceedingly valuable addition.

MASTERTON   NEW ZEALAND

Above: **Shorn wool is extremely bulky so needs to be compressed for transport. Before the days of the mechanical press, shed hands simply trampled the clip into bales. The earliest presses were introduced from Australia, but New Zealand also had its own, one being made by Donald & Sons of Masterton. In 1924 their No. 2 press provided for the sewing of the filled bale with a patented system of hinges, one of which was described as 'joggle-eyed'.**

68 A SHORT HISTORY OF SHEEP

Above: *The claimed advantage of Donald's No. 1 press was its portability, enabling it to be taken to the wool, instead of vice versa. Two men provided leverage for the rack-and-pinion system, and were able to deal to a bale in about six minutes. Like wool itself, Donald's presses were something of an export item, being especially popular in South America.*

Kaiapoi Woollen Manufacturing Company. Altogether it employed 600 hands, and its 80 looms operated night and day, consuming about 2500 bales of wool per year. The company's Kaiapoi mill was the first in the colony to be lit by electricity, and was said to be the largest and most complete in Australasia.

Other large woollen mills were operated by the Mosgiel Company, at Kaikorai and Mosgiel, employing 400 hands in all, while Ross and Glendining's Roslyn Mills at Dunedin employed 450 workers. Here Merino wool was converted into tweeds, plaids, flannels, shirtings, shawls, hosiery and mauds (travelling rugs). The Roslyn mill claimed to be the first south of the equator to produce fibre by the revolutionary new worsted process.

During the period 1868–1881, the number of sheep in Great Britain decreased by 7.7 million. This may have been a concern for the English farmer, but it could only be good news for his New Zealand counterpart. A *Cyclopaedia of Useful Knowledge* at this time reminded New Zealanders of the usefulness of sheep, an animal 'Endowed by nature with a peaceable and patient disposition, and a constitution capable of enduring the extremes of temperature, adapting itself readily to different climates, and thriving on a variety

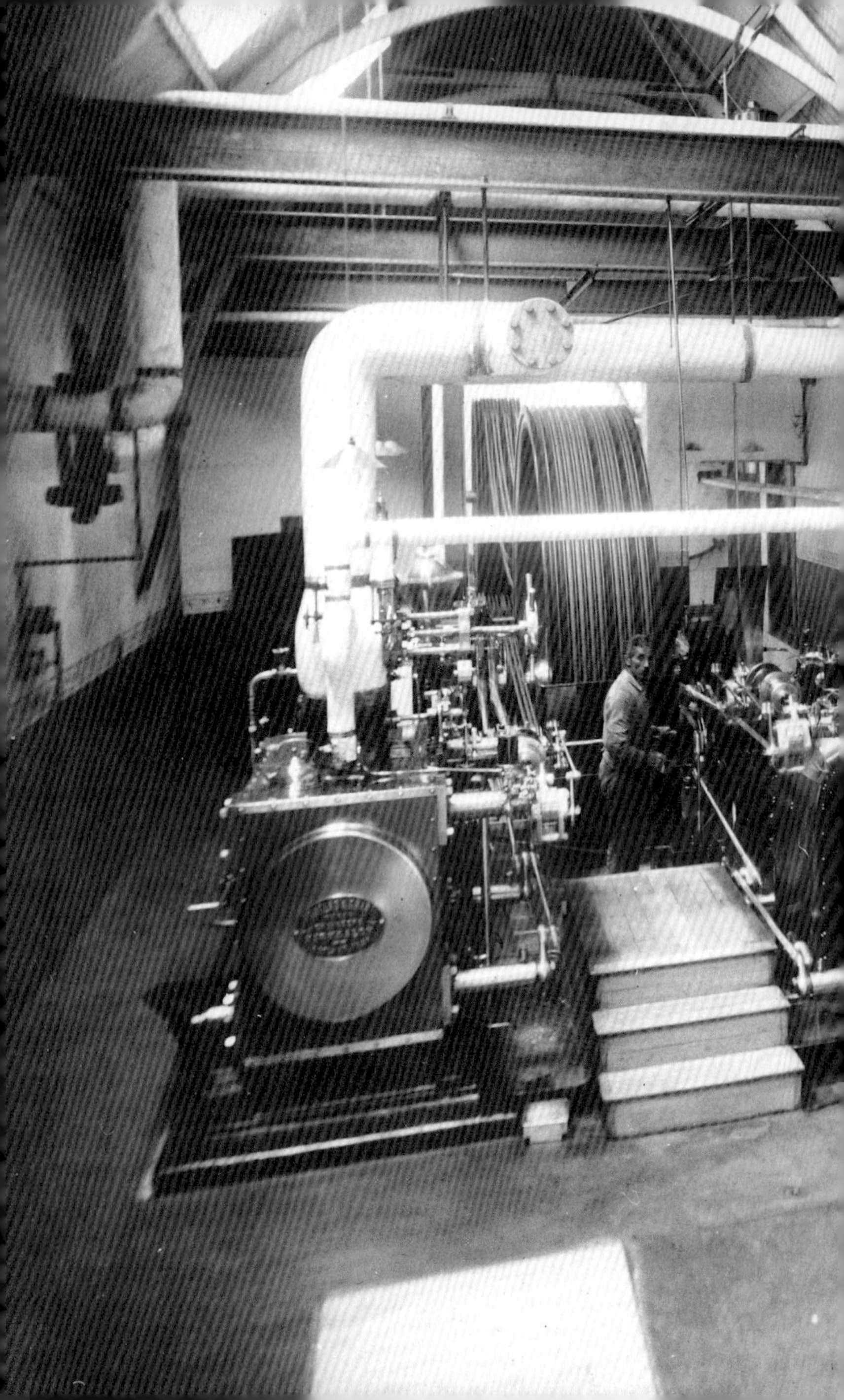

The engine room of the Mosgiel Woollen Mill, near Dunedin, around 1900. This business was the country's first proper woollen mill, and was established in 1871 after the local Provincial Council offered a bonus for the first 5000 yards of woollen cloth to be produced in Otago.

of pastures . . .' It had also 'afforded one of the most profitable and pleasing pursuits of man'.

The same authority described eight different breeds of sheep, including the Merino, whose wool was considered superior to all others. That sheep had dominated New Zealand's early flocks, satisfying the demand for wool. However it was not a profitable breed for mutton, which now became an important point in the light of the developments in 1882.

For anyone contemplating what appeared to be better prospects for farming in the colonies, there may have been no more colourful inducement to emigrate than that provided by *The New Zealand Handbook, or Guide to the Britain of the South* in 1879. It described how, 20 years earlier: '[Y]oung Squatters, sticking up a hut or two for themselves and shepherds, cleared a few acres for stock-yard and paddock; drew in their barrels of flour, chests of tea, bags of sugar, from the nearest store; took mutton from the plain, pig and pigeon from the bush; saw their flocks multiply on the wild herbage of the run, with little care of labour to themselves; sent down their annual dray-borne wool-harvest to their banker-merchant at the Port; rode into the Settlements as business and pleasure prompted; practised a rude but hearty hospitality among themselves; ever welcomed the stranger at their gates; and formed in New Zealand, as their fellows had long formed in Australia, a frank, jovial, high-spirited, pastoral aristocracy — a brotherhood of bronzed Bush centaurs, "Bearded like the Pard" [leopard], and creating, in the golden fleece, the Colony's best Export and most certain source of expanding wealth.'

Woolly words

DYED IN THE WOOL: Absolutely unchangeable. The wool of a black sheep will always remain that way, as no amount of dying can make it an acceptable white.

Above: A flock of Corriedales, the first sheep breed to be developed in New Zealand. It first appeared in 1868, and is now the fourth most populous breed in the country.

V

SHEEP BREEDS

Although the first sheep that came to New Zealand — the native Cape breed from South Africa, brought here by Captain Cook — didn't survive the experience, by the early nineteenth century the hardier Merino was well established in Australia, and would provide this country's first flock, landed here in 1834. Thanks to its wool and ability to adapt to local conditions, the Merino quickly became New Zealand's dominant sheep.

The introduction of refrigerated shipping then opened up new possibilities, and Merino were crossed with long-wool types to produce offspring with a heavier fleece and a better carcass.

By the early twentieth century, New Zealand considered itself 'admirably adapted' for all types of sheep. When it came to selecting suitable breeds, a *Settlers' Handbook* listed the 'dapper' Southdown and the 'more ponderous' Shropshires and Hampshire Downs, while English Leicesters were 'much sought after' for their ability to produce early-maturing lambs. The farmer also needed to take terrain into account: the Merino thrived on 'the wild lands of the colony' and on the higher country and the drier areas of the plains; while the Lincoln and Romney Marsh flourished on rich moist soils, and English and Border Leicester did well on the drier lands.

But the most profitable sheep could provide a combination of the best possible wool, the best carcass for freezing, and quick maturity. While the breed which best satisfied these requirements was a matter of opinion, one that was rapidly proving itself was of local derivation, and the result of the fourth cross of half-bred Lincoln–Merino and Lincoln rams: the Corriedale.

By the mid-1940s New Zealand had a sheep

population of 32.7 million, and the single largest breed was the Romney (6.1 million), followed by the Halfbred (2.2 million), Corriedale (1.1 million) and Merino (0.7 million), while there were nearly 22 million crossbreds. Sixty years on Romneys are still in the lead in terms of numbers, and New Zealand farms are home to over 30 different breeds. In addition there are some dozen feral breeds scattered throughout the country, including those on Chatham, Pitt and Campbell Islands. Unlike their domesticated relatives, they do not contribute to the national economy; in fact, they are more than likely to damage the native vegetation. But because these hermetic types have had little contact with modern breeds, they may carry important genetic characteristics and so may yet make a positive contribution to the wider sheep family.

Most of New Zealand's sheep breeds have originated from overseas — and some have only been here since 1990 — but this country has managed to develop a number of its own, beginning with the Corriedale. The main sheep breeds found here are listed in approximate order of decreasing numbers. It should be noted that just as New Zealand's sheep population altered dramatically from the mid-1980s, individual breeds have also been subject to considerable change as farmers adjust to the demands of the market and other variable factors.

New Zealand Romney
Originally from southeast England, where it is known as the Romney Marsh, this sheep arrived in New Zealand in 1853. In the early 1900s its numbers began to increase rapidly, and by the early 1960s it was found in all parts of New Zealand and accounted for three quarters of its 50 million sheep. The Romney has been crossed with the Cheviot to produce the Perendale, and with the Border Leicester to produce the Coopworth. The Romney is distinguished by its black nostrils and is known as a dual purpose sheep, having equal emphasis on meat and wool. The national Romney flock now numbers some 27 million.

Coopworth
This breed was developed in the 1960s from Border Leicester and Romney breeds, and its name honours the work of Professor Ian Coop of Lincoln College. There are now about 7.5 million Coopworths, and the breed has taken over from the Romney on wetter lowlands and easier hill country. This sheep is also dual purpose, and its fleece is used in the heavier apparel and carpet industries. A Coopworth ram has the

distinction of appearing on the coat of arms of former New Zealand Prime Minister, Sir Keith Holyoake.

Perendale
Registered in 1960, this was developed by Sir Geoffrey Perendale of Massey University by interbreeding Cheviot and Romney. The Perendale enjoys the hill country, and its wool has high insulation qualities which makes it in demand for blankets. Distinguished by its black nose, there are some 4.8 million Perendales in New Zealand.

Corriedale
This breed was pioneered in 1868 by James Little, the manager of Corriedale Station in North Otago, by mating British long-wool rams (mainly Lincoln and English Leicester) with Merino ewes and interbreeding the progeny. The intention was to produce a sheep that made better use of New Zealand's improved pastures, and resulted in this country's first indigenous breed. By 1924 it was also established as an export breed, and relocated in North and South America and Australia. The dual purpose Corriedale has a black nose, and there are some 2.7 million on the east coast of the South Island and the drier parts of the North Island.

The Lincoln School of Agriculture — later Lincoln College, Canterbury University, and now Lincoln University — opened in 1880 and is the oldest agricultural college in the southern hemisphere. The building seen here is Ivey Hall, which honours the first director, W. F. Ivey, who was also an early advocate for the use of artificial fertilisers, especially superphosphate.

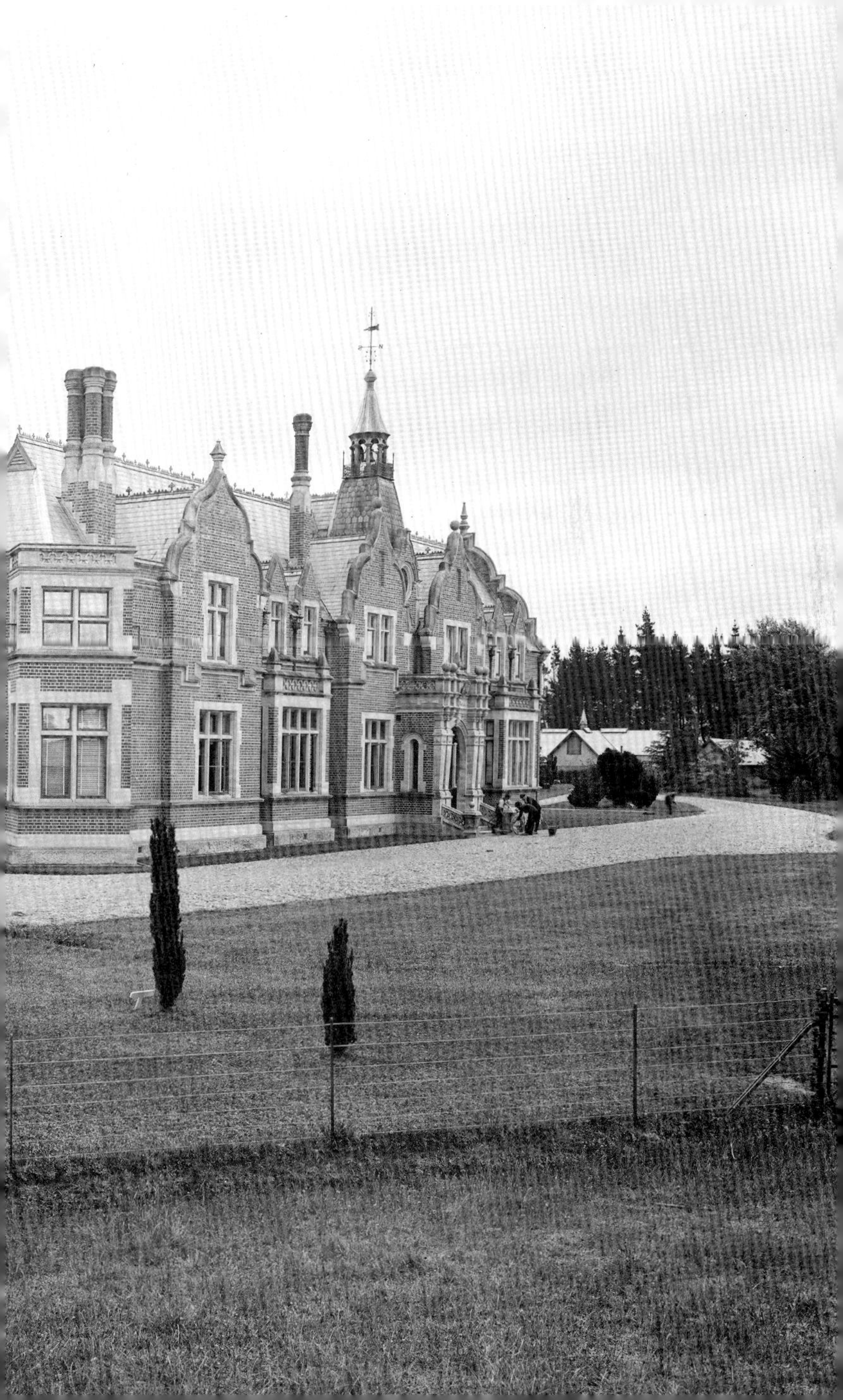

Merino

The oldest established and most numerous breed in the world, the Merino was first introduced to New Zealand in significant numbers in 1834. It is a slow-maturing specialist wool breed, and it remained this country's dominant breed until the early 1900s when it was replaced by dual purpose types such as the Romney, Border Leicester, English Leicester and Lincoln. The Merino provides the raw material for quality woollen and worsted fabrics. The ram is horned, and there are now some 2.5 million sheep on the high country of the South Island, the Canterbury Plains and some areas of the North Island.

New Zealand Halfbred

The nineteenth century need for improved meat and fertility characteristics resulted in the development of the Halfbred from the Merino and one or other of the long-woolled breeds — Leicester, Lincoln and Romney. There are now some 2 million of these sheep — distinguished by white face and pink nose — found mainly in the foothills of the Southern Alps.

Drysdale

This sheep resulted from research by Dr Francis Dry of Massey University in the 1930s and 1940s on genetics and the inheritance of hairiness. Developed from sheep of Romney or part-Cheviot origin, the Drysdale is a dual purpose breed, whose extremely long and strong wool plays an important part in locally made carpet blends. The breed has horns — the rams' are heavy while the ewes' are short — and there are some 600,000 throughout New Zealand.

Borderdale
In the 1930s the Border Leicester was crossed with the Corriedale to produce the New Zealand Borderdale. A dual purpose breed, it enjoys the fertile, dry areas of the South Island, and has a comparatively low susceptibility to footrot. Its wool is in demand from home spinners as well as commercial apparel manufacturers. The current Borderdale population is about 516,000.

Polwarth
Developed in Polwarth, Victoria, Australia, from Lincoln–Merino cross ewes and Merino rams, this breed came to New Zealand in 1932. A dual purpose breed able to adapt to a range of climatic and regional conditions, the Polwarth is found in Otago, Canterbury, Marlborough and Wairarapa. It has a pink nose and currently numbers about 160,000.

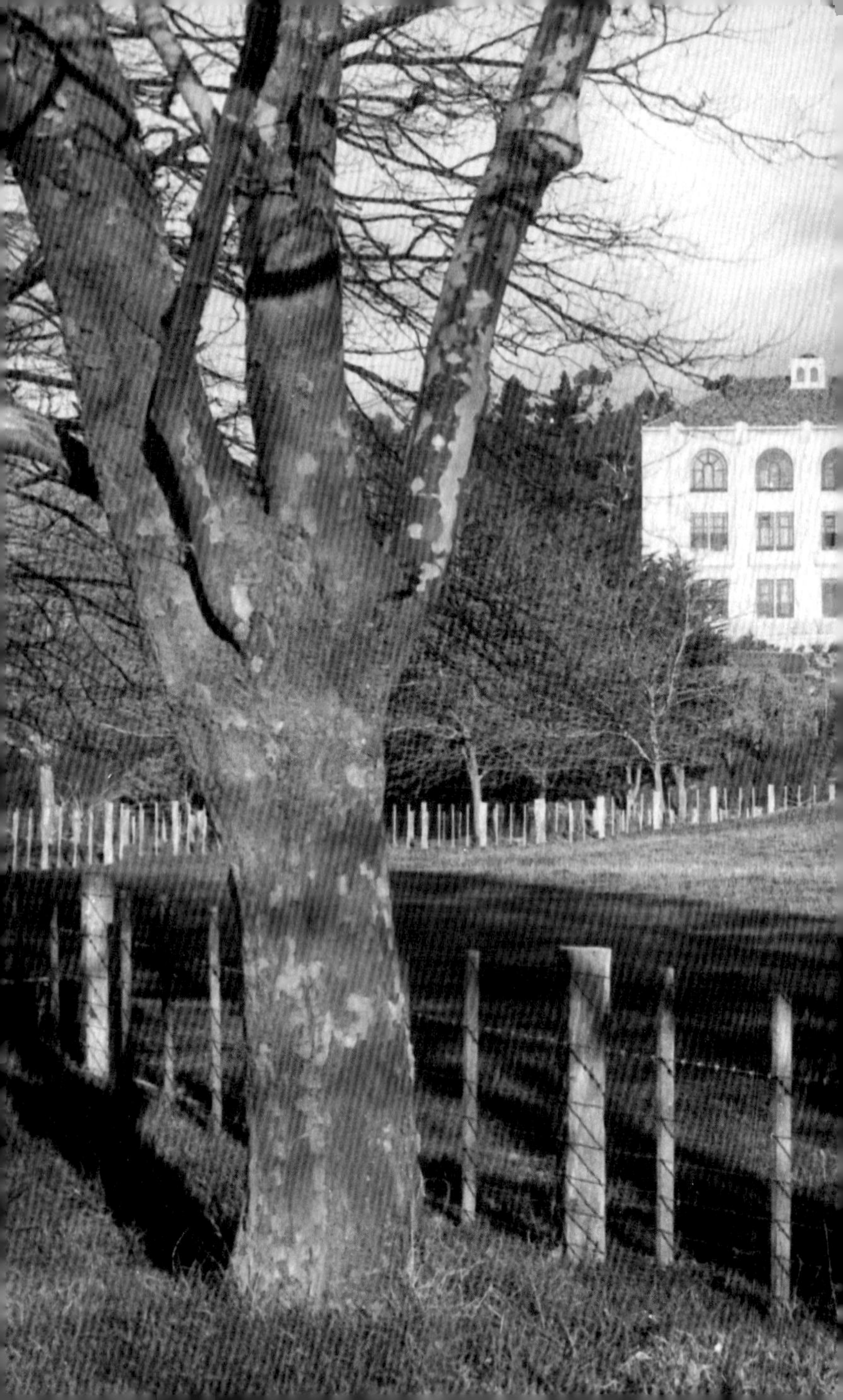

In 1926 New Zealand gained a second tertiary institution for studying farming practice with the founding of Massey Agricultural College (now Massey University) at Palmerston North. The first principal was Professor (later Sir Geoffrey) Peren, who was also responsible for developing the Perendale sheep breed.

Border Leicester
This sheep first arrived in New Zealand in 1859 and, following the introduction of refrigeration, was used to produce heavy-weight lamb and mutton offspring. It has been used to develop the Coopworth (Border–Romney cross) and Borderdale (Border–Corriedale cross), while the wool is used for upholstery and carpet yarns. The Border Leicester has what has been described as an 'ultra Roman nose'. There are some 110,000 Border Leicesters throughout the country, mostly in the lower half of the South Island.

South Suffolk
The overseas demand for leaner meat led to the development in Canterbury of the South Suffolk, from Southdown and Suffolk breeds. The new breed, combining quick maturity and high flesh-to-fat ratio, was registered in 1958. In addition to meat, the South Suffolk provides wool used for fine apparels and hand knitting yarns. The sheep has a dark brown face and ears, and there are about 94,000 of them throughout New Zealand.

Poll Dorset
Developed in Australia from the English Dorset Horn, the Poll Dorset was introduced to New

Above: *They may be under the watchful eye of a farmer, his daughter and a horse, but this small flock of Border Leicester ewes and lambs seem more interested in the photographer. This breed was introduced to New Zealand in 1859, and in addition to providing meat and wool it has been used to develop other breeds.*

Zealand in 1959. It is similar to the Dorset Horn, which it has almost replaced. The ewes are described as having 'feminine' characteristics, and the rams are able to breed over an extended mating season. The Poll Dorset is mainly a meat breed, and the skin is used in the fashion industry to make linings for boots and coats. There are some 85,000 spread around New Zealand.

Suffolk

The English Suffolk breed first arrived in New Zealand in 1913. A hardy sheep described as 'noble in appearance', the Suffolk provides lamb for export and high quality wool for hand knitting yarns, tweeds, flannels and dress fabrics. The sheep is distinguished by its black face and legs, and there are around 60,000 throughout New Zealand.

English Leicester

This was an early import to this country, arriving in 1843. It was used to produce sheep best suited to New Zealand conditions — Halfbreds from Merino ewes and fat lambs. Around 1900 the English Leicester was this country's third most popular breed — after the Merino and Lincoln — but was then overtaken by the rising populations of the Romney and Southdown. The breed's heavy curly wool is used for braids, suit linings, coatings and furnishing fabrics. There are now about 15,000 English Leicesters in New Zealand, mainly in the South Island and in Wairarapa and Hawkes Bay.

Dorset Down

Although it first came down to New Zealand in 1921, the Dorset Down soon disappeared, and today's sheep are the result of a re-introduction in 1947. It is an ideal sire for producing prime lamb, while the wool has been used in specialist industries. The Dorset Down has a brown face, ears and legs, and there is a population of some 15,000 scattered throughout New Zealand.

Texel
This sheep originated on Texel, a small island off the coast of Holland, and was made available in New Zealand in 1990. It is a hardy breed and a good forager, and also has the advantage of being dag free. It quickly contributed to the country's prime lamb trade, and the present population is around 15,000.

Wiltshire
This sheep was introduced to New Zealand from Australia in 1972 with the aim of siring fast-growing, large, lean lambs. Unusually, the sheep's wool is shed annually if left unshorn, and the sheep is free of the usual dags. Because it can survive in hot climates, the breed has also been an export item. The Wiltshire has black feet and nose, and there are about 15,000 throughout New Zealand.

Cheviot
Developed in the Cheviot Hills, between England and Scotland, this breed was introduced to New Zealand

in 1845. Large flocks were established in various parts of the country, but it went into decline from 1910. It was revived in the 1940s when crossed with the Romney, producing a hardier sheep that had increased growth rates and was easier to muster on hill country. This cross resulted in the new Perendale breed. Bulky Cheviot wool is used in carpets and knitwear, and there are about 11,000 in New Zealand.

Lincoln

The Lincoln is an ancient breed, and first came to this country in sizeable numbers when imported by the New Zealand and Australian Land Company in 1862. The sheep adapted readily to grazing on recently burnt-off forests, and it was second in popularity only to the Merino in the late nineteenth century. From the early 1900s it was gradually replaced by the Romney. It is now used mainly for producing crossbred ewes, while its strong wool has many specialist uses, such as in the wig industry. The Lincoln is a large and heavily built sheep, and some 10,000 are found throughout the country.

Finn

Also known as the Finnish Landrace or Finnsheep, this breed is, not surprisingly, a native of Finland. It

SHEEP BREEDS OF NEW ZEALAND

Border Leicester

English Leicester

Finn

Hampshire

Lincoln

Dorper

Dorset Down

Drysdale

East Friesian

Borderdale

Cheviot

Coopworth

Corriedale

South Dorset

South Suffolk

Southdown

Suffolk

Texel

Wiltshire

was introduced to New Zealand in 1990, and found to be particularly resistant to facial eczema. It is also recognised as a high fertility breed, and the ewe hoggets display what has been described as 'sexual precocity'. It's also a highly intelligent medium sized sheep with a short tail and a pink nose. There are only about 5000 Finns in New Zealand.

Southdown
The Southdown originated from Sussex, England, and first came to New Zealand in 1842. It was used in the evolution of the Hampshire Down, Suffolk and Dorset Down breeds. The New Zealand Southdown population increased rapidly after 1920 as a result of overseas demand for early maturing, lightweight lamb carcasses. In 1957, there were 786,000 Southdown ewes in New Zealand, but the sheep is now used for siring fat lambs, and it numbers around 5000.

Hampshire
The English Hampshire breed was developed by crossing Wiltshire and Berkshire ewes with Southdown rams and first came to New Zealand in the early 1860s. However, by the end of the century it had disappeared. It was reintroduced in 1951, and its main use is as a sire for the meat industry. The

Hampshire is a large sheep with dark brown face and legs. It is known as a very efficient converter of grass into meat, and there are fewer than 3000 breeding ewes in New Zealand.

East Friesian
Originally from the north of Holland and Germany, the East Friesian was released in New Zealand in 1996. It is noted for its high milk production, and its potential as a basis for an industry producing feta cheese and other sheep-milk products. There are about 1200 East Friesians in New Zealand.

Dorper
Developed in South Africa, this cross of Dorset Horn and Blackhead Persian was introduced to New Zealand in 2000. It is described as ideal for lifestyle blocks and organic farming, and it also makes an ideal pet. There are some 1000 purebred Dorpers in New Zealand.

Oxford Down
The Oxford Down was developed in England in the 1830s but was not available in New Zealand until 1990. This large sheep is used for siring meaty carcasses, and it also produces a large amount of wool. There are some 400 in stud flocks in New Zealand.

South Hampshire
The crossing of the Southdown and Hampshire in the 1950s led to the development of the South Hampshire, whose carcass has more meat and less fat.

Gotland Pelt
Originally from the Swedish island of Gotland in the Baltic Sea, this sheep produces high quality lightweight pelts of black or grey wool which are used in the quality garment industry.

Ryeland
A polled sheep — without horns — the Ryeland is one of Britain's oldest breeds, developed by monks in the fifteenth century. This sheep was introduced to New Zealand in 1903.

Shropshire
This breed arrived in New Zealand in 1864 and its numbers increased with the advent of refrigerated shipping. From 1885 to 1905 it competed with the Border Leicester as a popular fat lamb sire, but both were gradually replaced by the Southdown. The sheep has a brown face, ears and legs, and nowadays only small flocks remain in New Zealand.

South Dorset Down
The crossbred South Dorset Down resulted from the mating of Southdown ewes with Dorset Down rams, and became a popular prime lamb sire in the 1950s. Wool from slaughtered crossbred prime lambs sired by South Dorset Downs (and other Downs) is a major part of New Zealand's production of slipes — skins from which the wool has been chemically removed at the meatworks. The South Dorset Down has a brown face, ears and legs.

White Headed Marsh
This breed originated from the North Sea marshes of Germany and is similar to the Romney in appearance. It was released in New Zealand in 1990, and is used to inject size and higher lamb production into Romney and Perendale flocks. A large sheep with a black nose, its wool is principally used in the manufacture of carpets.

Black and Coloured Breeds
In what may be a legacy from their distant feral ancestors, some New Zealand sheep carry recessive genes responsible for fleece colouration. Occasionally such genes can coincide to produce a coloured sheep within a white flock. As a result of careful breeding, flocks of coloured breeds such as Romney (predominant), Coopworth, Perendale, English and Border Leicester,

Merino, Corriedale, Halfbred and Polwarth have been established in New Zealand. The colours of such sheep include shades of black, grey or brown, and vary from patterns or spots to full body cover. These flocks are mostly small and are used to produce wool for handcraft use, and such sheep are also a source of woolskin rugs.

In addition to the above, a number of other sheep breeds have appeared in New Zealand and failed to establish themselves. Back in the 1860s, the Cotswold and Dartmoor were bred here for a short time, while the fate of some Chinese rams said to have arrived at that time is unknown. The Wensleydale was imported in 1894, and did not survive a re-introduction in 1920, while other unsuccessful arrivals in the early 1900s were the Tunis (from America), the Roscommon (from England), the Scotch Blackface and, in 1937, the Kerry Hill from Wales.

Woolly words

BLACK SHEEP: An unpopular member of a group or family. On farms a black sheep — a throwback to the domesticated animal's feral ancestors — is generally unpopular, as a result of the preference for white wool.

Above: While various efforts were made to develop a canning process suitable for exporting mutton, all were superseded by the success of refrigerated shipping in 1882. Sheep continued to end up in tins, however; in the early 1900s the small print on the label of the Canterbury Meat Company's canned tongue assured consumers that the product contained not less than 14 grains of saltpetre to the pound.

VI

PRESERVING AND EXPORTING MEAT

By the 1880s all New Zealand sheep could at least anticipate being shorn annually and making a contribution towards the national wool clip. Some were also destined to satisfy the local demand for meat, while a more exclusive number could expect selected edible parts to be preserved for export. Otherwise, the large bulk of the sheep population was engaged in the (presumably) more pleasurable business of reproducing its own kind, to maintain a continuing supply of wool-producing units.

> ### Woolly words
> LEG O' MUTTON SLEEVES: Worn by fashionable women in the mid-nineteenth century, these sleeves were full at the shoulder and upper arm and fitted at the forearm.

As the size of the nation's flocks grew, so did the problem of what to do with excess or elderly sheep. Boiling down to extract fat and tallow for the soap and candle industries was the usual option, with the remainder of the animal destined to replenish the very pasture it had once enjoyed, as manure. The meat was of little use apart from supplying limited local consumption, but the potential of a huge market in Britain was an incentive for devising a reliable means of preservation for export.

In Australia, the canning of cooked beef had met with some success, but the processes had not yet been applied to mutton. In 1869, the price of wool was falling, and a group of runholders and businessmen in Christchurch met to discuss the best way of utilising surplus stock. The Canterbury Meat Export Company was formed, and decided to adopt an Australian process of preserving raw meat by dipping it in a solution of bisulphite of lime and then sealing it in tins.

Production was initially delayed by a shortage of suitable materials and qualified tinsmiths, but by 31 May, 10,044 sheep had been dealt to and converted into 160,120 lbs (72,630 kg) of tinned meat. In early July, a shipment of 842 cases sailed for London on the *Caduceus*, spurred on by a Canterbury Provincial Government offer of £1000 for the first successful meat export industry. The company survived for about five years, but like others who adopted similar approaches, it experienced difficulties — including defective tins — and eventually closed down. It was obvious that canning technology held little promise for a meat export industry, but a more likely approach was suggested when a shipment of chilled beef was successfully sent from the United States to Britain in 1876.

The first attempt at sending a frozen shipment of mutton and beef from this part of the world was in 1876, from Australia, on the *Northam*. But the refrigeration machinery broke down during the voyage, and the venture was a failure. However, success came in 1879 when a cargo of frozen Australian beef and mutton reached London intact on the *Strathleven*. It was soon followed by another consignment of 4600 sheep and lamb carcasses and 100 tons of butter. The second shipment had left from new freezing works in

Melbourne. These works were inspected by Thomas Brydone of the New Zealand and Australia Land Company, with a view to establishing such an industry on the other side of the Tasman.

New beginnings

The pioneer of the frozen meat industry in New Zealand was Edinburgh-born William Saltau Davidson, who arrived in this country in late 1865. He went to work as a shepherd at the Levels Station in South Canterbury, which a decade earlier had been visited by the legendary drover Mackenzie. That run now had a flock of over 85,000 Merinos, and Davidson set about developing a new pure half-breed more suited to pastures of imported English grasses, crossing Merino ewes with imported Lincoln stud rams.

In 1878 he returned to Scotland and became general manager of the New Zealand and Australia Land Company. Two years later, following the success of the two cargoes of Australian frozen meat to Britain, Davidson recognised the possibilities for New Zealand. He approached the Albion Shipping Company, and as a result one of its best and fastest sailing ships, the *Dunedin*, a barque-rigged iron ship of about 1250 tons, was fitted out with insulated meat chambers, boilers and refrigeration machinery. Meanwhile, back in New Zealand, Thomas Brydone arranged the preparation of sheep for the first shipment.

A killing shed was erected on the company's Totara Estate, some ten kilometres southwest of Oamaru, and the services of first-rate butchers were secured. The *Dunedin* arrived at Port Chalmers, and each morning 240 carcasses were sent to the ship by train, packed in special vans cooled with large blocks of ice.

William Davidson had come back to New Zealand on the *Dunedin*, and on 7 December 1881 he and Brydone personally began the stowing on board of the first frozen sheep ever loaded in New Zealand. Things proceeded well until a fracture in the engine's crankcase stopped work, and necessitated the sale of the 641 sheep already packed in icy chambers between decks, as well as the 360 others killed and on their way to the wharves. As a result it was New Zealanders, not Britons, who got to eat what would have been an historic load of frozen meat. Repairs were carried out and the *Dunedin* finally sailed on 15 February 1882. Along with mutton, lamb and beef, the ship also carried an extremely mixed cargo of pheasants, hares, rabbits, turkeys, geese, ducks, chickens, fish, butter, milk and eggs.

Although there were no problems with the

Above: The Dunedin *leaving Port Chalmers on its historic voyage in February 1882. The success of the first cargo of refrigerated meat sent from New Zealand to England both changed the type of sheep raised in this country, and stimulated the construction of a large number of freezing works.*

refrigeration machinery during the voyage, there were some anxious moments when sparks from the funnel (of the boiler) set fire to the sails on several occasions. There was added drama when a tropical hurricane tore spars away and stoved in all the ship's boats. Also, when the ship was in the tropics it was discovered that the cold air was not circulating as intended, and so the captain personally crawled into the main chamber to cut extra openings to ease the situation. In the process he became so numb from the cold that the mate had to rescue him by tying a rope to his legs and hauling him out from behind.

Originally, about 60 passengers had booked passage on the *Dunedin* but, because of fears there might be a recurrence of mechanical problems mid-ocean, all but two opted out. One who stayed was a seventeen-year-old who soon became aware of the benefits heralded by that historic voyage, and observed how pleased the crew were to be fed fresh mutton instead of their traditional 'salted junk' and bully beef. When the *Dunedin* arrived safely off the Isle of Wight, directors of the shipping company — now Shaw, Savill & Albion — came aboard and were also able to enjoy some of the fresh-tasting frozen fare from New Zealand.

The ship reached London Docks after a voyage of 98 days. The precious cargo, which according to one source amounted to 4311 carcasses of mutton, 598 of lamb, 22 pork and 2226 sheeps' tongues, arrived in good condition, and the only casualty was one carcass which had been 'bumped'. The shipment was sent to London's Smithfield Market, and was all sold within a fortnight, with mutton and lamb fetching 6½ pence per lb — a profit of 3¼ pence per lb. Although some of the sheep had been frozen for over four months, it looked like freshly killed mutton, and the shipment even received a favourable mention in the House of Lords.

While the *Dunedin* was on its way to London, a second ship, the *Mataura*, also equipped with refrigerating machinery, arrived at Port Chalmers for a similar undertaking. The total number of sheep carcasses loaded and frozen on board was 3844, of which 496 were provided by John Grigg, one of the great pioneer farmers of Canterbury. He was founder of Longbeach Station, near Ashburton, which would shortly carry upwards of 40,000 sheep. The *Mataura*'s refrigerated cargo was even more of a menagerie than the pioneering *Dunedin*'s. It included six bullocks and 77 suckling pigs, 26 hares, 24 rabbits, 66 fowls, 24 barracudas, 18 ducks, eight hapuka, three pukeko, two cases of cheese, 12 cases of hams and one cask of penguin skins. It all arrived safely, but the prices fetched were not so high, and freezing operations that same year. This operation began with a novel approach, using the hulk of the ship *Jubilee*, which lay at the company's wharf and contained sufficient refrigerating machinery to deal with 600 carcasses of mutton a day. It could also be towed alongside ocean-going ships in the harbour, so that meat could be transferred from one freezing chamber to another with minimal exposure to the air.

The refrigerated meat trade represented a valuable new export for New Zealand, which no doubt explains why the Gear Company chose a frozen sheep carcass as the centrepiece of its display at the 1885 New Zealand Industrial Exhibition in Wellington. For reasons of visitor comfort, this representative of the colony's new economic future was changed daily. But the Gear Company was still keeping its options open, also displaying no less than 33 varieties of its canned preserved meats. They included smoked ox- and sheep-tongues, jugged hare, venison and fish, while the list of customers for these was equally varied, including both the British Admiralty and the German Navy.

Within nine years of the *Dunedin*'s voyage there were seventeen freezing works in New Zealand. By 1889 the number of carcasses exported exceeded one million, and in the first half century of operation the number sent to Britain was some 200 million. At first this new

Thanks to refrigeration, New Zealand was able to cater to the growing market for mutton and lamb in Britain. This impressive display, including carcasses of prime lamb from the Auckland Farmers Freezing Company, awaits customers in London in the early 1900s.

605
The BRITISH NEW ZEALA
605 B.N.Z.
B.N.Z. 605
AUCKLAND FARMERS FREEZING Co LTD. -N.Z.-
A. ORR

this experience set the standard for future sheep exports to London.

> **Woolly words**
> MUTTON CHOPS: The name for luxuriant side-whiskers popular with Victorian gentlemen was inspired by a cut of meat.

The works

The *Dunedin* has been described by historian Gavin McLean as 'The ship that changed our destiny'. In all, it made ten trips carrying frozen goods to London before it met its own refrigerated destiny in 1890, probably after striking an iceberg off Cape Horn. Following the success of these early shipments, freezing works were established in this country. The first to go into operation was the New Zealand Refrigeration Company at Burnside (Dunedin) in 1882. Next to be formed was the Canterbury Frozen Meat and Dairy Produce Co Ltd, whose board of directors included John Grigg and whose Belfast Freezing Works opened in February 1883.

A third pioneer was The Gear Meat Preserving and Freezing Company in Wellington, which began

Above: Following the success of the Dunedin *shipment, William Nelson established a freezing works at Tomoana, near Hastings, in 1883–34. In its first season it shipped 40,000 carcasses. Nelson was also involved in other works at Ocean Beach in Southland, Sockburn in Canterbury, and Gisborne. In 1924 the original Tomoana works (seen here) was rebuilt, and was considered one of the most up-to-date in the country.*

era in the retail meat trade was dominated by mutton, but by 1900 lamb was firmly established, reflecting customer tastes in Britain and bringing necessary changes to the farms of New Zealand.

Within a little more than a decade of the voyage of the *Dunedin*, the most important industry in New Zealand based on the value of manufactures or produce was its meat works. In 1894 the country had 43 establishments dedicated to the freezing, preserving and boiling down of animals, and employed a total of 1568 hands. Sheep were the main source of raw materials, and also contributed to the colony's second most important manufacturing industry, tanning, fell mongering and wool scouring, which engaged 1196 hands. In tenth place were eight woollen mills, providing work for another 1175 people, while nineteen soap and candle works around the country dealt with another byproduct of the sheep.

When the frozen meat trade was initiated, there was some debate as to whether New Zealand would be able to provide one million sheep each year without depleting its breeding stocks. But within a decade the number of carcasses exported had reached 1,607,754, and at the same time the national flock had increased by just over 800,000 to 19.38 million. Sheep farming accounted for 59 per cent of the country's exports (in descending order of value: wool, mutton, tallow, sheepskins and preserved meats), but it was believed the country was carrying only a little more than half the number of sheep of which it was capable. By opening up additional land, more farmers would be able to exploit what the 1894 *New Zealand Official Year-Book* termed 'that most valuable of animals — the sheep'. The outlook was also extremely promising, for it anticipated that exports could be increased fourfold.

> ## Woolly words
> LAMBS TO THE SLAUGHTER: Innocent and helpless, unaware of the imminent danger — as in sheep being trucked off to the freezing works. 'I will bring them down like lambs to the slaughter, like rams with the goats.' (Jeremiah 51:40)

The Southern Frozen Meat and Produce Export Company was founded in 1882 and erected its first works (pictured) at Bluff. It continued operating until 1912, when the works was shifted to Makarewa, north of Invercargill. In 1890 a second freezing works began near Bluff, at Ocean Beach. This was a private venture by local businessman Joseph Ward, who became Liberal Prime Minister of New Zealand in 1906.

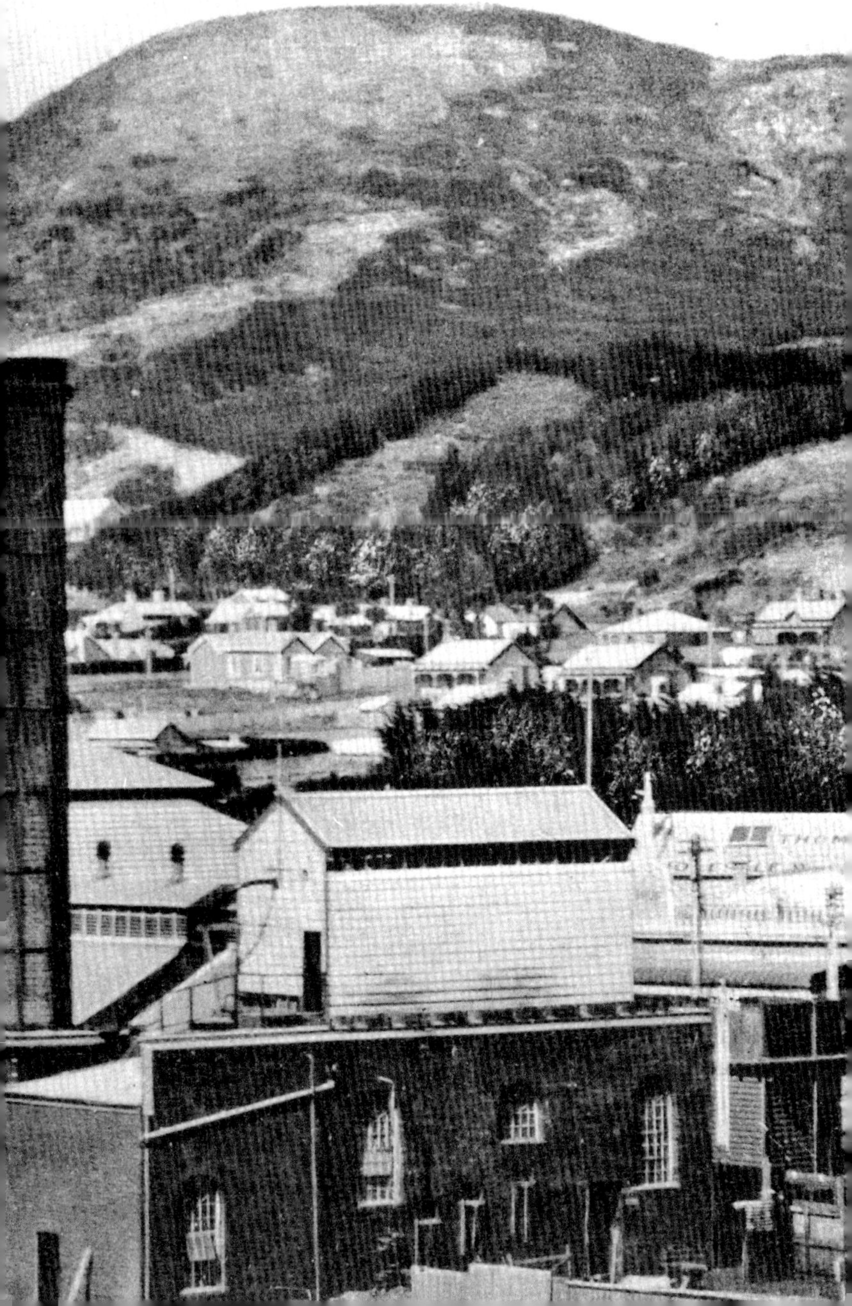

A taste for lamb

Thanks to its climate and soil, it was believed New Zealand was the most suitable of Britain's colonial possessions for sheep farming. In fact, it felt even more blessed for that purpose than the mother country itself, as it was spared long and severe winters and such diseases as foot and mouth. While New Zealand was the main source of imports of mutton and lamb to Britain, it was only a small proportion of that country's total meat consumption. Home production was responsible for about 82 per cent of all meat consumed in Britain and New Zealand contributed two per cent.

The good news for New Zealand farmers was that Britain's own production of mutton couldn't keep up with population growth, so a greater dependence on imports seemed likely. Twenty years earlier, such meat had been largely the preserve of the upper and middle classes, but the standard of living of the working classes — representing two-thirds of the total population of 37.7 million — was on the rise, and they were developing a taste for New Zealand lamb.

Although Australia had begun shipping frozen mutton and lamb to Britain three years before the *Dunedin* set sail, New Zealand had quickly caught up and exceeded its neighbour's offerings. By 1893 this country had supplied 11.3 million carcasses, which was a little over half the total quantity of frozen mutton and lamb imported into the United Kingdom since the trade began. The only other significant contributor was Argentina, which provided about two-fifths, while Australia's portion was one-tenth. There were now 36 ships — including six sailing vessels — engaged in transporting New Zealand meat to consumers in Britain.

In the year to 30 April 1893, New Zealand killed 1.9 million sheep for export and another million for local consumption, out of a total of nearly 19.4 million. Six million of the latter were Merino ewes, which were crossed with Downs and long-woolled types to produce 'freezers' suitable for export. The Merinos stayed behind, for while their mutton was not highly regarded in Britain, it was acceptable here. At that stage Canterbury was the source of far more sheep meat exports than any other part of this country, providing just over one quarter of all mutton carcasses and one half of all lamb carcasses. The province became synonymous with sheep, and so New Zealand lamb and mutton came to be known in Britain as 'Canterbury lamb', while the term 'prime Canterbury' was also used.

As for the most suitable breeds of sheep, one

authority favoured the Border Leicester cross as the most profitable all round. In England, the Home Leicester was produced mainly for the carcass, but crossing the Border Leicester ram with the Merino ewe produced a sheep which both fattened early and had a profitable fleece. And by introducing Down and Leicester breeds, some North Island farmers were said to be producing 'Wellington' mutton that was fetching the same price as 'Canterbury'.

> **Woolly words**
> GRASS ROOTS: The basic source, or the ordinary people.

Pastures new

Small farming now became an economic proposition. Previously, the economic situation and the importance of wool as the major export had favoured large pastoral holdings. But there was now increasing pressure on land, and in 1890 the new Liberal Government introduced its policy of encouraging closer settlement by making Crown land available to genuine farmers.

In 1892 an Act enabled the Government to buy

PRESERVING AND EXPORTING MEAT 117

Above: A careful inspection of lamb and sheep carcasses at the Canterbury Frozen Meat Company, around 1910. This company, the second of its type to be formed in New Zealand, began in 1881 and was initiated by a group including John Grigg of Longbeach. By the 1890s the company's success at promoting Canterbury mutton and lamb on the London market had led to supplementing the original freezing works at Belfast, north of Christchurch, with others at Fairfield, near Ashburton, and Pareora, near Timaru.

private land for such purposes, and the first such major purchase was the 84,755 acre (34,325 ha) Cheviot Estate in the rolling hill country of North Canterbury. Some 80 people and about 80,000 sheep had occupied this estate, and it was now bought (for £260,000), subdivided and opened up with roads. Within a year it was home to nearly 900 people and 74,000 sheep, as well as other livestock.

One who farmed here after drawing a property in a ballot was George Forbes, who was the son of a ship chandler from Lyttelton and who later became Prime Minister of New Zealand. The township that served this new farming community was first known as McKenzie, in honour of the Liberal Government Minister of Lands, but it later reverted to the name of the district's original station, Cheviot.

The development of Cheviot set the pattern for the policy of cutting up large freehold sheep runs into medium-sized arable and livestock farms on plains and lowlands, and into grazing runs on steeper land. In the period 1892–1911 the Government purchased 209 large estates, totalling 1.2 million acres (486,000 ha), which were subdivided into 4800 holdings. At the same time it encouraged the settlement and establishment of farms in bush-covered areas of the North Island.

Above: *Mansion House on the Cheviot Estate, around 1890. Wealthy landowner William Robinson had named the property after the Cheviot Hills, on the border between England and Scotland. In 1893, the year after Robinson's death, the Liberal government purchased the estate and subdivided it into smaller allotments to encourage settlement.*

This increase in the area of sown grassland saw dramatic changes to both the sheep population and its distribution. Total numbers rose from 12 million in 1881 to 16.7 million in 1891, and then to 19.3 million in 1899. And while the South Island had always been ahead, that year saw the North Island take over, slightly, and by 1911 it was home to 53 per cent of the nation's flock.

A scientific approach to farming was encouraged with the foundation of the Canterbury Agricultural College at Lincoln, in 1880, while the Department of Agriculture was established two years later. Another important development at this time was the introduction of the topdressing of grasslands, which probably first occurred in the Waikato region. Sheep were now able to enjoy pastures nourished with fertilisers rich in the essential elements of phosphorus and nitrogen, derived from the droppings of birds.

Woolly words
MUTTONBIRD: The sooty shearwater, known to Maori as titi. An officer of the Royal Marines allegedly once referred to petrels on Norfolk Island as 'flying sheep', and so they became know as muttonbirds.

PRESERVING AND EXPORTING MEAT 121

Above: **Outbuildings, around 1910, on the Mt Peel Station, one of several established in 1855 by John Acland and Chales Tripp. When their partnership was dissolved in 1862, Acland took over the Mt Peel property.**

Above: Sheep undergoing their annul dip on the Adkin farm, Levin, in 1906. Farmhands with long-handled dipping crutches ensure adequate immersion.

VII

HOW TO LOOK AFTER SHEEP

Thanks to a temperate climate and extensive grasslands, sheep enjoy a relatively high standard of living in New Zealand. And as a result of its isolation, this country's border controls and quarantine systems have kept sheep free of many serious diseases found elsewhere.

The welfare of New Zealand's sheep was considered as early as 1884 when a Police Offences Act provided for fines and imprisonment for the ill-treating, over-driving, overloading, abusing or torturing of animals. Owners were required to provide sufficient food, water and shelter; while no slaughtering, branding, conveying or carrying was to subject animals to unnecessary pain or suffering. From 1893, Stock Acts regulated aspects of animal management, and guarded against the spread of disease.

The dipping of long-woolled and crossbred sheep was made compulsory once a year (between 1 January and 31 March in the North Island, and between 1 February and 30 April in the South Island), and the fine for failure to do so varied from three pence to two shillings per undipped sheep. If the animals were found to be affected with lice or ticks, an Agricultural Department inspector could order an immediate dipping (for the animals), or another fine might ensue. Immediately after shearing, all sheep were to be branded with a wool-mark, an identification that was to be not less than three inches long. This could be of pitch, tar, paint, raddle (red ochre) or lamp-black, mixed with oil or tallow. Sheep also had to have one of the following: an ear mark, a metal clip affixed to the ear, a tattoo mark on the skin, or a fire-mark on

HOW TO LOOK AFTER SHEEP 125

Above: There is no indication of the active ingredient of this sheep dip, but other such products in the early 1900s contained nicotine in the battle against lice, tick and scab.

the horn or face. Ear marks were made by cutting, splitting or punching, but the removal of more than one-quarter of the ear could result in another fine, of up to £10 per sheep.

Around 1895 New Zealand began to emerge from a long depression. While farmers enjoyed improving prices for mutton and wool, from 1910 they also had the Agriculture Department's new publication, the *New Zealand Journal of Agriculture*, to keep them abreast of technical and scientific developments. Soon it would be possible to gain a university degree in agriculture in this country, and in 1927 Massey Agricultural College (later Massey University, and named after Prime Minister William Massey) was established at Palmerston North. But New Zealand was about to feel the effects of another worldwide depression, and woolgrowers were hit particularly hard.

As far as New Zealand sheep were concerned, the 1930s began with the introduction of the Drysdale breed and ended with the re-establishment of the Cheviot. South Island farmers faced the invasion of their pastures by South American Nasella tussock, but science had now found a cure for cobalt deficiency in sheep and cattle. The economy picked up as the decade progressed, but the Depression had encouraged farmers to make do, where possible, with

HOW TO LOOK AFTER SHEEP ♈ **127**

Above: A paddock of swedes: an important supplementary fodder crop for sheep, particularly in southern New Zealand. The vegetables can be eaten in the field, or collected and fed out as required. In the 1950s New Zealand sheep could snack on swede varieties with such names as Superlative, Grandmaster, Crimson King, Resistant and — perhaps the tastiest of all — Sensation.

materials at hand. *The Weekly News* offered a selection of handy hints to save time, money and labour, and this catalogue of Kiwi ingenuity included numerous useful ideas for looking after sheep.

> ### Woolly words
> LIKE LOST SHEEP: Wandering around aimlessly. According to the parable, if a shepherd had 100 sheep and one went astray, he ought to leave the 99 and go into the mountains in search of the lost animal. Great would be the rejoicing when the lost animal was returned safely to the fold.

Do it yourself

If a farmer, for example, needed to brand wool bales and had run out of black paint or tar, a suitable substitute could be made by mixing soot with linseed oil. Rural recycling included converting old benzine tins and 40-gallon drums into feeders for lambs or salt-lick containers, while a cunning method of mixing sheep dip was to pour the essential chemicals down a 7 ft (2.13 m) length of down-spouting, while stirring vigorously. This technique was said to ensure a more evenly mixed dip, and doing so eliminated the risk of poisoning the first sheep to take the plunge.

One of the more drastic suggestions was a dosing device. The farmer plugged one end of an eight-inch (20 cm) length of iron piping with a cork, and placed in it the sheep's required dosage of pills. He then inserted the open end of the pipe into the patient's mouth and, when it was about three inches (76 mm) down the throat, he lifted the device so that the pills slid down and were securely administered. 'Shepherd' of Wairoa knew of stations where this method had been used 'for years' and, with sheep comfort in mind, he advised that the end of the pipe needed to be smooth.

Woolly words
RED SKY AT NIGHT: Said to be a shepherd's delight, suggesting tomorrow would bring favourable weather for looking after sheep.

Any farmers needing to manhandle a sheep in the 1930s may have appreciated a technique recommended by a *Weekly News* reader from New

Above: An attentive dog checks the tyres before a truckload of sheep leaves a Wainuiomata farm, northeast of Wellington, in 1938.

Plymouth: 'If standing on the left of the sheep put the left hand on its nose and with the right hand take hold of the left hind leg. At the same time twist the head to the right and pull the hind leg up under the belly and the sheep will be sitting at your feet. If standing on the right of the sheep, reverse the action.'

As well as all this improvisation, the 1930s sheep farmer could use conventional proprietary cures and devices. Black Leaf 40, containing 40 per cent nicotine sulphate, controlled stomach worms, lice, ticks and scab in sheep, but elsewhere around the farm this versatile product could deal to insect pests in the garden and lice in the fowl run.

Another annual task was made easier by the Daroux Emasculator, which guaranteed bloodless castration for lambs, rams and calves. It could be operated with one hand, and a happy customer from Waimate had used it for three seasons on a 'drop' of some 1800 lambs, finding it as simple and reliable as the 'old knife' method.

Japan's entry into the Second World War in 1941 cut New Zealand's access to the rock phosphate from Nauru and the nearby Ocean Island that was used in the fertiliser industry. Five years later, when supplies resumed, experiments began with aerial topdressing to improve soil fertility. Back on the ground, from the

1960s onwards farmers were replacing their horses with motorbikes for getting round their properties.

Heading and huntaways

Farmers still depended heavily on one other animal, the working dog, and in 1966 there were more than 200,000 of them in the country, and most of them helped farmers with their 50 million sheep and 6.5 million cattle. The population of dogs also included some which specialised in rabbiting, so in a sense they were also in the service of sheep.

New Zealand sheep have stimulated two more unusual forms of sport: shearing competitions and sheepdog trialling. The latter had begun in South Canterbury in 1889, with a national organisation formed in 1957.

Standard competitions involve tasks similar to those the shepherd and the dog encounter during the course of normal farm duties, and are in two classes: heading and huntaway. The first requires the dog to round up sheep and bring them back to his master, and in the other canine challenge the sheep are worked upwards and away from the master. The original shepherding dogs brought to this country were Collies, who worked silently. But early New Zealand shepherds soon found the need for a noisy dog to flush sheep out from bush cover, and the huntaway was bred for this duty.

One of the points of the annual muster is to submit sheep to the dip. Farmers in the mid-1950s were advised against dipping in extremes of heat or cold, or in wet weather, and also not to do so late in the day so that the sheep were reasonably dry by nightfall. Rams needed particular care and were to be dipped two to three months before 'going out' with the ewes; any closer to 'tupping' and they were in too spirited a state. Rams also needed extra care in the dip, as they were not natural swimmers.

There was a controversy in 1985 when the New Zealand Government allowed the export of live sheep, a trade strongly opposed by animal welfare groups. By the end of the decade, more than one million live sheep were being exported each year — mainly to Saudi Arabia, which does not produce enough of its own for both sacrificial purposes during religious observances, and domestic consumption. Following an unacceptably high mortality rate on a 1990 shipment, the trade to that country was suspended. It recommenced in 1991, and a code of welfare was introduced to maintain standards. Since then the trade in live sheep to Saudi Arabia has tapered off to about one shipment per year, involving some 40,000 animals.

New Zealand sheep can now expect certain minimum standards of animal husbandry. In 1996, the Ministry of Agriculture and Fisheries produced a code with the animal's welfare in mind, pointing out that sheep have a number of special requirements. They should not be deprived of feed for longer than 24 hours, while overfeeding can also cause problems such as obesity, which can result in woolly sheep becoming cast.

Sheep pant in hot weather, and the evaporation of moisture from their nose and mouth increases water requirements. Newly shorn sheep need to be provided with shade to prevent sunburn, and may also require up to 40 per cent more feed for some three weeks after shearing to maintain condition. And while sheep are obviously used to moving in flocks, they can become distressed if left alone or isolated from their own kind.

Woolly words
SHEPHERDING: An illegal tactic, as on the rugby field, whereby the player with the ball is shielded by a team mate from a would-be tackler.

Above: *With a bit of encouragement from behind, sheep are mustered across a bridge in hilly country at Huiarua, inland from Tōkomaru Bay, on the east coast of the North Island.*

Above: New Zealand sheep enjoying hay, a supplementary feed preferred in higher rainfall districts. Further south, and where winter conditions are severe, bulky root and forage crops such as swedes and oats are used.

Code of care

The 1996 code of care suggested lambs' tails should be docked before the age of six weeks, and by means of a rubber ring or a hot searing iron. And whereas young rams were once subjected to an emasculator, they can now expect the slightly more friendly sounding elastrator, which effects castration by applying a rubber ring to the neck of the scrotum. But this process should only be carried out if absolutely necessary, as there is no 'ram taint' in the meat of young lambs.

As for general comfort, dags are an attraction for blowflies and need to be trimmed, as should wool growing around the sheep's eyes, which can obscure vision. Recognising the sensitivity of ear tissue, the marking of sheep by removing a portion of the ear is not now recommended, unless there is no practical alternative. But if it is carried out, the amount removed is not to exceed one fifth of the ear, which was a little less than the maximum permitted in 1893.

In their first year sheep are particularly susceptible to internal parasites and generally require regular dosing with a drench gun, and vaccination now prevents such diseases as pulpy kidney, blackleg, black disease and tetanus. Animals with footrot or

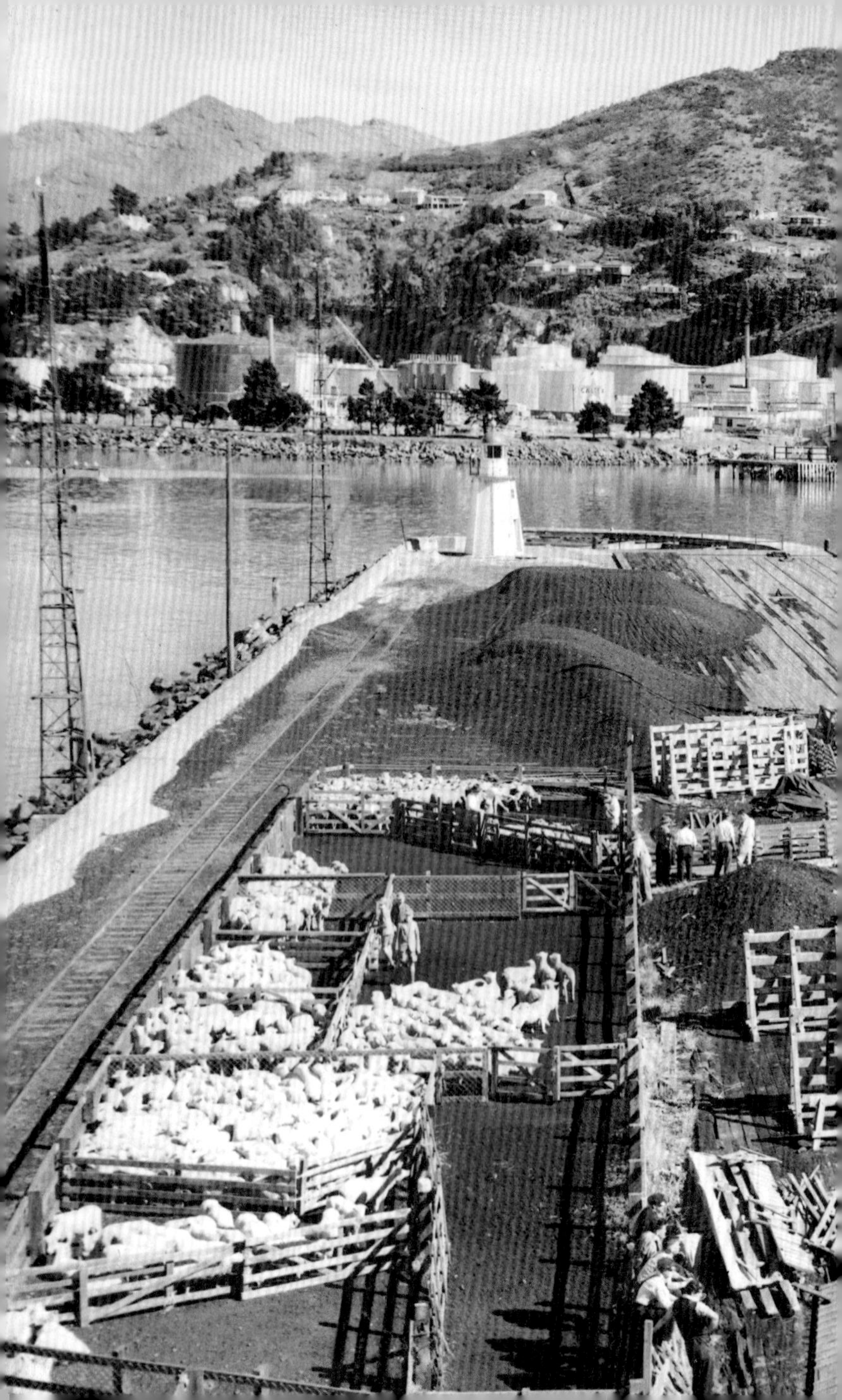

While the vast proportion of New Zealand's sheep exports have been as frozen carcasses, live animals have also been sent overseas. These Corriedales on the wharf at Lyttelton, around 1950, were about to be shipped to Kenya on the Thala Dan.

footscald (benign footrot) should be stood in a solution of zinc sulphate or formalin, as opposed to the old cure of carbolic acid or caustic.

Fungus in the pasture is the cause of several diseases, including facial eczema, which occurs in much of New Zealand, apart from the lower South Island. Affected sheep need to be moved to pastures with lower spore counts.

The staggers is caused by a fungus in ryegrass and produces tremors and a staggery gait in sheep, necessitating them to be moved safely away from any natural hazards. Merino wethers, having already been subjected to the elastrator, are vulnerable to a bacterial infection known as pizzle rot. Rich pastures can cause urine to have a high ammonia content, which can scald the animal's foreskin, causing discomfort and increasing the risk of flystrike. Fortunately, it can be prevented by a minor surgical procedure, which facilitates urine drainage.

But the ramifications of the animal welfare code go far beyond the paddock and sheep run. New Zealand's economy depends on its access to international markets, and that in turn demands quality assurance, of which the well-being of our sheep is a large and vital part.

Woolly words

DAG: Strictly speaking, a piece of wool clotted with dung that hangs about the hindquarters of sheep. Such accretions are an attraction to flies, and need to be removed. They have also been associated with humour, a dag being a funny person. Such a character can also be 'a bit of a dag' or just plain 'daggy'. The dimension of sound is included in the expression to 'rattle one's dags', meaning to get a bit of a hurry-on.

*Above: **The durable qualities of woollen serge and saddle tweed fabrics would have been well known to these farmers attending the Addington saleyards, Christchurch, in 1951.***

VIII

TWENTIETH CENTURY SHEEP

On 26 September 1907 New Zealand graduated from a Colony to a Dominion. Although it was little more than a name change, the growing sense of maturity in the nation received a further boost four years later with the granting of our first national heraldic bearings, or coat of arms. A national competition for a suitable design gave a fair indication of what New Zealanders felt was appropriate.

Suggestions for the supporters — the pair of figures on either side of the arm's central shield — reflected various aspects of the young Dominion's industries and history, and nominated candidates included farmers, soldiers, Maori warriors and even the extinct moa.

The competition was judged by two cabinet members, with input from the Curator of the National Museum in Wellington. The winning design featured a European woman (Zealandia) holding the national flag and a Maori warrior wielding a taiaha, the pair supporting a shield emblazoned with symbols of the New Zealand economy. Its maritime dependence was represented by three sailing ships, its geological resources by two mining hammers, and agriculture and farming by a wheat sheaf and — naturally — a fleece.

New Zealand thus officially recognised its huge economic debt to the sheep. It was not an entirely original idea, for the state of New South Wales had placed a golden fleece — along with wheat sheaves — on its coat of arms back in 1906. Later, some of New Zealand's larger cities reflected the regional dependence on sheep; the coats of arms of Wellington, Napier and Christchurch also incorporated fleeces, while that of Hastings had a ram supporter (on

Above: *New Zealand's first Coat of Arms, granted in 1911, was updated in 1956 (seen here). The changes included turning the two supporters (Zealandia and the Maori warrior) to face one another, and standing them on fern leaves instead of golden curlicues. The modified arms continued to acknowledge the country's debt to the sheep, in the form of a fleece on the central shield.*

the dexter side), and those of Dunedin, Invercargill and Wanganui each had a horned ram's head. A Coopworth ram also served (as it were) as supporter (sinister side), along with an Aberdeen Angus bull (dexter side), on the coat of arms granted in 1979 to

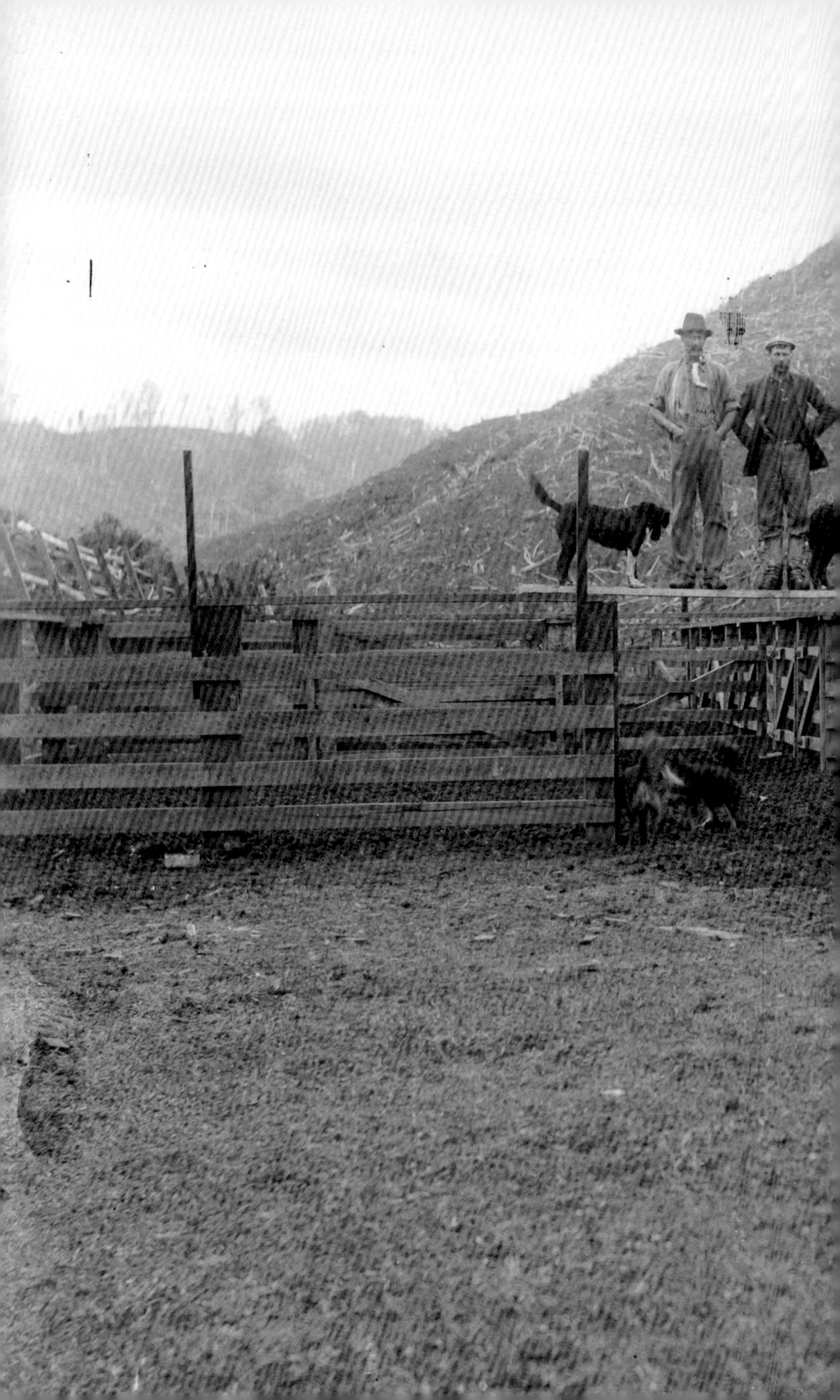

Newly constructed saleyards, ready to receive sheep to populate the hills behind. The landscape is scarred by remnants of recently destroyed bush, making way for pastures. A typical scene, probably inland Taranaki, 1907.

Sir Keith Holyoake (1904–84), Prime Minister and Governor General of New Zealand. Both the ram and the bull were animals bred by Holyoake, who had been a farmer, while a small kiwi on the crest of his arms acknowledged that he was also known as 'Kiwi Keith'.

In the early decades of the twentieth century, New Zealand sheep farming expanded as virgin land was brought into production, and sown grasslands were improved through the use of better pasture mixtures and fertilisers. Sheep owners may have felt their interests were being looked after by William 'Farmer Bill' Massey, Prime Minister from 1912 to 1925. Born in Ireland, he had arrived in New Zealand in 1870. He gained his first local farming experience with his father (who had emigrated earlier) at Tamaki, Auckland, and with John Grigg at Longbeach, Canterbury. He became president of the Mangere Farmers' Club before moving into politics, and later led the wartime coalition National Government and two Reform Governments.

During Massey's ministry, the Discharged Soldier Settlement Act of 1915 enabled returned soldiers to be placed on land made available by the Crown. About 9500 men were financed on to farms by this scheme but, when produce prices fell in the early

Above: Although sheep products are acknowledged as important contributors to the 'sinews of war', it is the dairy cow that predominates on this patriotic promotion from the 1940s.

1920s, many found themselves in financial difficulties and some had to walk off their land. A similar scheme for the rehabilitation of ex-servicemen followed the Second World War, and by 1961 had settled a total 12,179 men on farms, either developed by the State or purchased privately.

Growing flocks

As pastures were both expanded in quantity and improved in quality, land that was previously able to support only dry sheep could now take breeding ewes. From 1900 to 1945 the nation's flock increased from just over 19 million to 33.42 million. The largest proportion of these were breeding ewes (61.5 per cent), followed by lambs (26.5 per cent), wethers (7.5 per cent) and dry ewes (2.7 cent). Rams, charged with responsibility for maintaining numbers, represented just 1.6 per cent.

Two thirds of the nation's sheep were now crossbreds — nearly 22 million — followed by Romneys (6.15 million), New Zealand Halfbreds (2.18 million) and Corriedales (1.1 million), while the once dominant Merino was relegated to fifth place (718,000). The North Island continued to be home to more sheep than the South Island, while individual land districts with the most sheep in 1947 were Wellington (including Wairarapa, 6.7 million), Canterbury (5.5 million), Hawke's Bay (4.3 million), Otago (4 million) and Southland (3.4 million).

The North Island could also claim the larger number of flocks — 16,577 compared to 16,300 in the South Island. The nation now had only 15 flocks numbering over 20,000 sheep, while smaller flocks were likely to be found also on dairy and mixed farms, either taking advantage of land not suitable for cows, or with the two ruminants happily occupying the same pastures.

In 1940–1 New Zealand celebrated a century of British sovereignty. The sheep had been in this country for a little longer, and a publication in a series of Centennial Surveys offered an interesting perspective on its contribution to the nation.

In *New Zealand Now*, Oliver Duff detected a struggle between 'two great families of domestic animals' — sheep and cows. He identified two social systems, a result of there being sheep on one side (east) of the North Island and cows on the other.

In his view, sheep 'make' gentlemen while cows 'unmake' them; the one leaving farmers with clean hands and feet and giving them freedom, while the other dragged them into the mud and enslaved them.

Above: *A small proportion of the nation's flock in the early 1950s. By then New Zealand's total sheep population was about 35 million, and two-thirds of them were crossbreds.*

Sheep were responsible for creating a 'gentlemen class' who enjoyed the easy life, as opposed to west coast dairy farmers who worked seven days a week.

When *New Zealand Now* went to a second edition in 1956, Duff conceded his original vision of national types was now outdated, for the policeman, soldier and public servant had all changed. Only the farmer was 'more or less' the same, although now 'alarmed rather than elated' at the fabulous wool prices received for the previous five seasons.

Woolly words

FOLD: An enclosure for sheep, or pen. Also a flock of sheep and, figuratively, the members of a church. The best known use with regard to sheep is in Byron's poem 'The Destruction of Sennacherib':

> 'The Assyrian came down like
> a wolf on the fold,
> And his cohorts were gleaming
> in purple and gold.'

A stable market

Whatever the sheep farmer felt about his wool receipts, he could take comfort from a guaranteed return, thanks to market intervention. In 1916 the Crown had become the sole buyer and seller of the nation's wool, and for a period Britain bought all we could produce, and at a fixed price. At the outbreak of the Second World War, Britain once again agreed to buy all New Zealand's output, and by the end of hostilities it had stockpiled over 10 million bales, including some 1.75 million from this country. An organisation was set up to dispose of these without upsetting the market, and managed to do so by 1951.

The Crown was superseded as a wool purchaser by the New Zealand Wool Commission, whose mission was to ensure minimum or 'floor' prices at wool auctions, which had resumed following the end of bulk purchasing by Britain. In 1950, the prospect of another war — in Korea — and a subsequent huge hike in world wool prices led to the establishment of a Wool Proceeds Retention Fund, with a proportion of sale receipts 'frozen' to control the boom.

In 1951 the waterfront dispute, the costliest in New Zealand's industrial history, disrupted a number of shipments of wool exports which were not able to take advantage of the higher prices. The buying and

Above: *Female employers at the Westfield freezing works, Otahuhu, 1943. Meat was then in short supply as a result of wartime rationing, with preference going to New Zealand troops serving overseas. The rationing of meat, along with tea and sugar, was finally abolished in 1948.*

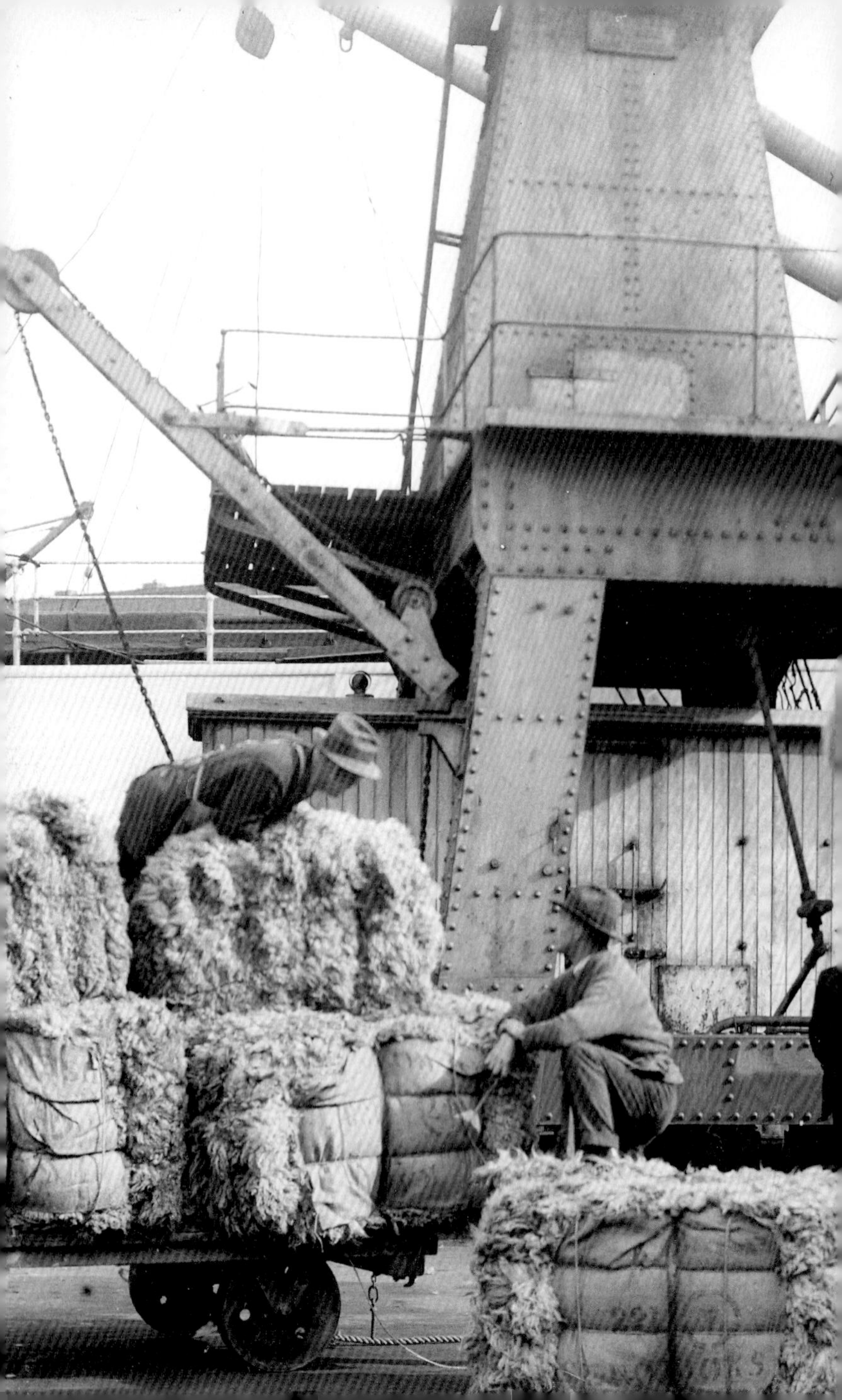

As well as maintaining the national wool clip, sheep can make one further contribution to the economy. Here, bales of sheep skins await export from the Wellington wharf in the 1920s.

selling of New Zealand wool by statutory marketing bodies continued, with the aim of minimising price fluctuations and providing guaranteed returns to growers. Meanwhile, the nation's stockpile continued to grow, and when all forms of market intervention ended in 1991 it had reached a peak of 655,000 bales. This mountain was then disposed of, and the sale of the last bale (weighing 123 kg) on 20 December 1995 marked the end of an era.

The export meat trade encouraged the development of two types of farms, for sheep with two rather different purposes in life — and in death. On fattening farms, which exploited highly fertile land, rams of the fat lamb breeds were put to ewes, and all their progeny were slaughtered. These ewes were brought in for breeding, for one or two seasons or the duration of their breeding life, and then they too were sold for slaughter. None of the ewe lambs was retained for breeding, so fat lamb farmers replaced their ewe stocks by buying in from breeding flocks operating on less fertile hill country, where the sheep were mostly Romneys.

The mid-1950s was a period of steady prosperity for New Zealand. But in 1958 there was a decline in export prices, along with another development that would soon have massive implications for the nation's

TWENTIETH CENTURY SHEEP 159

Above: Building a firm foundation for New Zealand's stockpile of wool, in the 1940s. It reached a mountainous 655,000 bales in 1991.

sheep farmers — the signing of the Treaty of Rome. This established the basis of the European Economic Community, or 'Common Market'.

New Zealand farmers were now at least able to take advantage of new advances in veterinary science. Intestinal worms and other such afflictions, once controlled by nicotine and carbolic, were now attacked by antibiotics, while the arrival of the first broad spectrum drench, Thibenzole, in 1962 heralded a new era in animal welfare, enabling farmers to run more sheep per acre.

By the mid-1960s New Zealand had about 50 million sheep tended by about 41,000 owners, and nearly half of its total land area of 66 million acres (26.73 million ha) were given over to sheep. The country was the world's largest exporter of mutton and lamb, and only Australia exported more wool. A major proportion of those products still ended up in the one place — Britain — which took 43.3 per cent of all our exports.

But during this period the New Zealand farmer was experiencing a decline in income relative to other sectors of the community, as a result of several factors, including increasing interest rates, wages, killing charges, transport costs and land values. The government introduced various measures to assist the rural sector, beginning in 1972 with the sheep retention and farm income equalisation schemes. There was a further challenge, although it hardly came as a surprise when Britain finally signed up with the European Community in 1974 and New Zealand lost its traditional market and faced the prospect of international competition.

> **Woolly words**
> PULLING THE WOOL (OVER SOMEONE'S EYES): In the sixteenth and seventeenth centuries fashionable men of wealth wore wigs, whether they were bald or not. A person could be fooled by having his wig tilted, thereby preventing him from seeing what was really going on.

Industry subsidies

The following year's measures to help farmers cope with the change included a $50 million meat and wool income stabilisation scheme, the suspension of meat inspection fees, and other supplementary payments. But more help was sought, and in July

1976 National Prime Minister and Minister of Finance Robert Muldoon announced the Livestock Incentive Scheme, designed to increase livestock numbers.

Two years later the Supplementary Minimum Prices (SMP) scheme guaranteed a minimum amount paid to farmers for meat, wool and dairy products. Cash grants, low-interest loans and tax rebates were also offered, and various government charges were removed to assist in such areas as purchasing farms and providing for such essentials as fencing, drainage and fertiliser.

By the early 1980s, subsidies were said to be providing some New Zealand farmers with up to 40 per cent of their income. Freemarketeers now considered such reactive measures as unsustainable, and as disincentives for innovation in the farming industry. They felt these protections needed to be removed to allow the economy to become more rational and efficient, even if it involved a painful period of reconstruction. Meanwhile, the incentives remained in place and New Zealand's stock numbers increased, the sheep population rising to reach an all time high of 70.3 million on 30 June 1982.

Ironically, this apogee was attained exactly a century after the event that had stimulated the sheep industry

Above: Sheep are herded in single file down a race to the farmer operating a drafting gate. He directs his flock into different pens, according to age, condition, or perhaps suitability for the freezing works. Drafting should also sort out the occasional feral goat which found itself rounded up in the muster.

in the first place, the voyage of the *Dunedin*. As the skyrocketing sheep numbers suggest, subsidies and tax incentives favoured quantity rather than quality, and offered no encouragement for diversification from traditional agricultural practices and products.

In August 1983 Opposition finance spokesman Roger Douglas offered a predictably grim perspective on the New Zealand economy. Since 1955 the gross national product per capita had slipped from third to about twentieth place among the Organisation for Economic Co-operation and Development (OECD) nations. What's more, it had fallen faster than any other such country in the process. Douglas used wool as a measure of our indebtedness: in 1975 it took 283 bales to pay the interest bill on New Zealand's borrowings, and by 1982 it took a staggering 1,400,000. For the current year he estimated 2,000,000 would be required.

He stated that since 1975 New Zealand society had been 'progressively corrupted, divided, and humiliated', with drug dealers and property speculators appearing to be the only people doing well. He put part of the blame on the Government's policies of subsidies and ad hoc controls and, in 1984, as Minister of Finance in the fourth Labour Government, he got his chance to do something about the situation.

> **Woolly words**
> TO FLEECE: To strip or rob of money or property. Akin to a wild animal removing the fleece from a sheep.

Rogernomics

In his budget of 8 November 1984, Roger Douglas outlined measures designed to ensure that some key prices in the economy reflected more accurately the true costs of production. In the area of agriculture, various assistance schemes would be lowered or removed, with some measures taking effect immediately.

'Rogernomics', the name given to the process of taking away the artificial props that had insulated the New Zealand economy from that of the wider world, proved a painful process for some. On 30 October 1985 a group of mid-Canterbury farmers protested at the Government's policies by slaughtering 2500 ewes. They objected to the low price then being paid for cull stock, but also drew attention to the difficulties facing farmers generally, most of whom were experiencing a downturn in income.

In all, nearly 30 different agricultural subsidies were

Above: *By 1975 the New Zealand sheepfarmer — and his dog — were likely to be getting around the flock on a farm bike, the internal combustion engine having superseded the original form of horse power.*

abolished, but the industry survived, by consolidating and becoming more efficient. One dramatic effect was a decrease in the national flock, down to 60.6 million by June 1989. Nearly half now consisted of Romneys (27.7 million), followed by 7.5 million Coopworths, 4.8 million Perendales, 2.7 million Corriedales, and 2.4 million Merinos, while another 10 million were classified 'unspecified' and 3 million were 'other' breeds.

Within six years sheep numbers had plunged another 12.5 million, down to 48.1 million in mid-1995.

While the ranks of both sheep and cattle were rapidly dwindling, the first animal to be domesticated — the goat — was now enjoying increased numbers on the farms of New Zealand. By mid-1985 there were about 283,500 of them, kept for their milk, mohair and meat, as well as the first function they had performed for man — weed control.

> **Woolly words**
> BELLWETHER: Dates from a Middle English practice of using a castrated ram with a bell around its neck to lead a flock of sheep. More recently, a bellwether is a region with political tendencies which are an indication of the bigger picture, electing the party which (in a democratic system) wins the overall vote. Hamilton East and Hamilton West are said to be 'bellwether' or indicator seats in the New Zealand electorate.

As the sheep population dropped, its centre was also shifting. After nearly a century of having fewer sheep, in 1996 the South Island reclaimed the lead, now with 26.85 million sheep to the North Island's 22.52

Above: In the late 1800s the Gear Meat Company supplied canned goods to the British Admiralty. This nautical connection was maintained over half a century later, in 1956, when the company's Petone works received a visit from the Duke of Edinburgh, who served in the Royal Navy during the Second World War.

million. Deregulation demanded diversification, and there were now other changes taking place on farms. Apart from rabbits, sheep once had the countryside largely to themselves, but they were now being required to share. While properties exclusively farming sheep represented the largest single use of New Zealand's grassland, they were now outnumbered by dairy farms, while a large number of farms now combined sheep with beef and other activities such as cropping.

The United Kingdom was once the only market New Zealand needed, but by 1995 it accounted for only 9 per cent of our total wool exports, behind Hong Kong and China. Both the export volumes and the real prices had declined, mostly as a result of competition from synthetics such as nylon, rayon, polyamides, polyesters and acrylics. However prices have improved for strong (crossbred) wool, of which New Zealand is the world's largest producer, which is used mainly in interior textiles such as carpets, upholstery, furnishings and rugs.

By the mid 1990s, Britain remained a major market for both New Zealand's lamb and mutton, followed by Germany. Other countries had their preferences: France, Belgium/Luxembourg, Japan and Saudi Arabia for lamb; and South Korea and France for mutton. Back home, the average New Zealander's annual meat consumption was down slightly, with an increase in chicken and pork at the expense of mutton. Just as sheep were used to sharing their paddocks with other animals, New Zealanders were, for now, enjoying alternatives to their traditional cuts of meat.

Above: A young shearer deals to a rather startled sheep with a pair of blades, in the early 1900s.

IX

SHEARING

Missionary John King may have started something in the early 1820s when he took to his sheep at the Bay of Islands with a pair of scissors. By then wool growing was already a well established industry in New South Wales, and more effective means of removing fleeces were introduced to this country.

Shearing was the most important event of the wool grower's year. At first it was a summer job, but the season was later extended and some sheep could anticipate two annual trims. In the early years, the job was done outside, but the building of woolsheds allowed large numbers of sheep to be held overnight to be kept dry for an early morning start. Shearers did the job with blades, in short scissor-like movements, and this technique predominated until the first shearing machines were introduced to New Zealand from the late 1880s. Longer strokes were then developed, increasing the speed with which a sheep could be shorn and enabling gangs to keep up with the growing number of sheep and the demand for their wool.

Shearers soon devised a suitable attire for the shed. Naturally, woollen singlets and trousers were ideal for absorbing perspiration and preventing shearers from catching chills, while improvised moccasins made from pieces of old wool pack stitched up with twine were both soft on the feet and didn't slip on greasy floor boards. Worn-out wool packs had other uses; for example, one of the suggestions from the *Weekly News* in the 1930s was that with a bit of trimming and sewing they could be turned into stock covers for other animals on the farm.

Shearing was extremely thirsty work, and 'smoke-ho'

Above: Various shearing techniques are shown in this view of a shed from the Illustrated New Zealand News *of 24 December 1883. Presumably the preferred approach is being demonstrated by the pipe-smoking 'Boss' at bottom right.*

4013.

A trio of New Zealand shearers takes to the outdoors, sometime in the early 1900s. The worker on the left appears to be picking up handy hints from his more confident colleagues.

provided a welcome break. Freshly brewed tea and perhaps scones were often sent up to the shed from the farmhouse, and the *Weekly News* again suggested just the sort of container needed for carrying the refreshments — an old soap or candle box with a handle made of a piece of pipe or hollow bone over a length of the ubiquitous number eight wire.

> ### Woolly words
> BULLSWOOL: Absolute nonsense or a total exaggeration. Obviously, bulls don't grow wool. Similar in meaning to bullshit, which bulls do produce in substantial quantities.

Early shearing machinery was often powered by steam or, later, electricity but more isolated areas used petrol- or diesel-fired generators. A lengthy rotating shaft ran above the 'stands', or shearers' positions on the boards, and provided the motive power for their handpieces, with such well-known brand names as Cooper ('double-ribbed driving wheel with flanged edge for strength'), Wolseley ('no running expenses except 6 pence a day for current') and Lister.

The importance of wool was illustrated on 21

SHEARING 177

Parts for Lister "No. 3" Shear

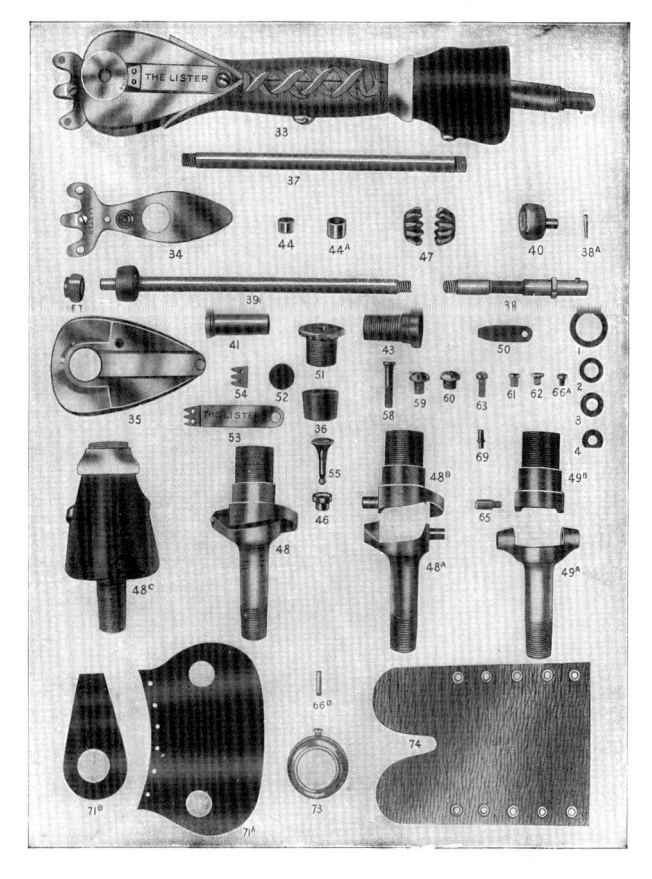

Above: Some of the vital components of the Lister No. 3 handpiece in the 1920s, powered by electricity or a diesel engine.

December 1916 when New Zealand's entire clip for that season was requisitioned for war purposes. All wool growers were required to deliver their bales to the stores of designated government wool brokers at one of thirteen ports around the country. Farmers on the east coast of the North Island had the choice of Tokomaru Bay, a coastal port until the early 1920s, or Gisborne.

All delivered wool was then examined, classified and valued according to prices received for the 1913–1914 season, to which the Government added 55 per cent. On receipt of payment, the bales became the property of His Majesty, and any wool owner who failed to follow these provisions was liable to have his wool seized. The wool was valued according to twelve different categories, ranging from superior Merino combings (up to 14 ½ pence per pound) down to inferior crossbred (from 7¾ pence per pound).

The Bowen style

Wool was again considered a vital material during the Second World War, and harvesting it was considered essential work. One shearer who was exempted from military service so that he could continue trimming the national flock was Walter Godfrey Bowen, born in Hastings in 1922. He and his brother Ivan developed a more refined and continuous shearing technique, eliminating unnecessary movements, and their improved method became known as the 'Bowen style'.

One of their innovations was using the non-shearing hand to stretch out the sheep's skin, producing a fleece that was more evenly shorn and therefore more valuable.

The new approach was demonstrated in 1953 when Godfrey Bowen, using a Wolesley number ten handpiece, sheared 456 ewes in nine hours at Opiki, Manawatu, and set a world record. He, along with Ivan (who later beat his brother's record), were soon in demand on the New Zealand Agricultural and Pastoral (A&P) Societies' show circuit. Not surprisingly, one or other of the pair won most of the shearing contests they entered — or at least came second to his brother.

The New Zealand Wool Board was also quick to recognise Godfrey Bowen's talent, both for his remarkable shearing technique and his value as a great promoter and advocate for this country's wool overseas. In 1953 the Board appointed him chief instructor of its shearing section, and he formed a team of instructors and established courses at Massey and Lincoln Agricultural Colleges.

Above: From a sheep's point of view, one advantage of being shorn by Godfrey Bowen was that it wouldn't take long. This subject was dealt to in February 1953, the year Bowen set a world record for shearing 456 sheep, in full wool, in nine hours.

> **Woolly words**
> SHEEPISH: Bashful, embarrassed, over-modest, diffident or timorous. It is not clear whether sheep are inclined to feel this way, but they may appear to do so on account of keeping their heads down, as they are invariably eating.

While early New Zealand shearers no doubt added excitement to their task by comparing individual performances in the woolshed, the first appearance of shearing as a competitive sport in this country may have been at the Hawke's Bay A&P Association in 1902. There were competitions during the 1930s and in conjunction with annual shows following the Second World War, but the idea of a national championship contest was first considered by the Wairarapa Young Farmers' Club in Masterton in 1957.

A shearing competition held in conjunction with that year's local A&P Show at Masterton proved popular, leading to the formation of the Golden Shears International Shearing Championship Society in 1960. Godfrey Bowen helped organise the first event, which was held the following year in Masterton and attended by nearly 300 shearers from New Zealand and Australia. Bowen also competed, coming second to brother Ivan, who won the open championship in the first year.

Godfrey Bowen gave shearing demonstration tours around the world, his extensive itinerary including the Soviet Union, where in 1963 he was honoured by Nikita Krushchev. Seven years later he directed New Zealand's main outdoor display at Expo '70 in Japan, and this led to the development of the Agrodome, on a 160 ha sheep farm near Rotorua. With local farmer George Harford, Bowen devised and established a theme park dedicated to the New Zealand farm. Bowen died in 1994, and that popular tourist venture is now run by the two founders' sons. Among its attractions are a working farm with 1200 sheep — along with other animals — where the Agrodome offers demonstrations by shearers and of working huntaway dogs, and includes the recreation of a shearing shed from the 1950s.

A true profession

In the mid-1900s a novice shearer could feel a certain sense of accomplishment by managing a tally of 100 mature sheep in a day. Half a century later that figure had risen to at least 200, a result of a major transformation in the job. Shearing's profile had been raised, for now it was considered a profession — and perhaps even an art form. It had a new sense of importance, and this image makeover was the result of the efforts of the Bowen brothers.

In 1961 there were about 7000 regular shearers in New Zealand. Godfrey Bowen was hardly a 'regular', but he was certainly the fastest. He sheared 559 small light-woolled sheep in nine hours with the usual meal spells, in Wales, breaking the previous world record of 501 Cheviot–Romney cross ewes, set at Okaihau, Northland. Shearing records are difficult to compare because of varying conditions, but in 1961 at Puketitiri, Hawke's Bay, Bowen raised his New Zealand record to 463 sheep (average size Cheviot-cross, each with about 7lb or 3.1 kg of wool) shorn in nine hours.

Woolly words

WOOLLIES: Garments made of wool, as in winter woollies. Woollies was also the name given to Woolworths, when it operated a chain of convenience stores around New Zealand.

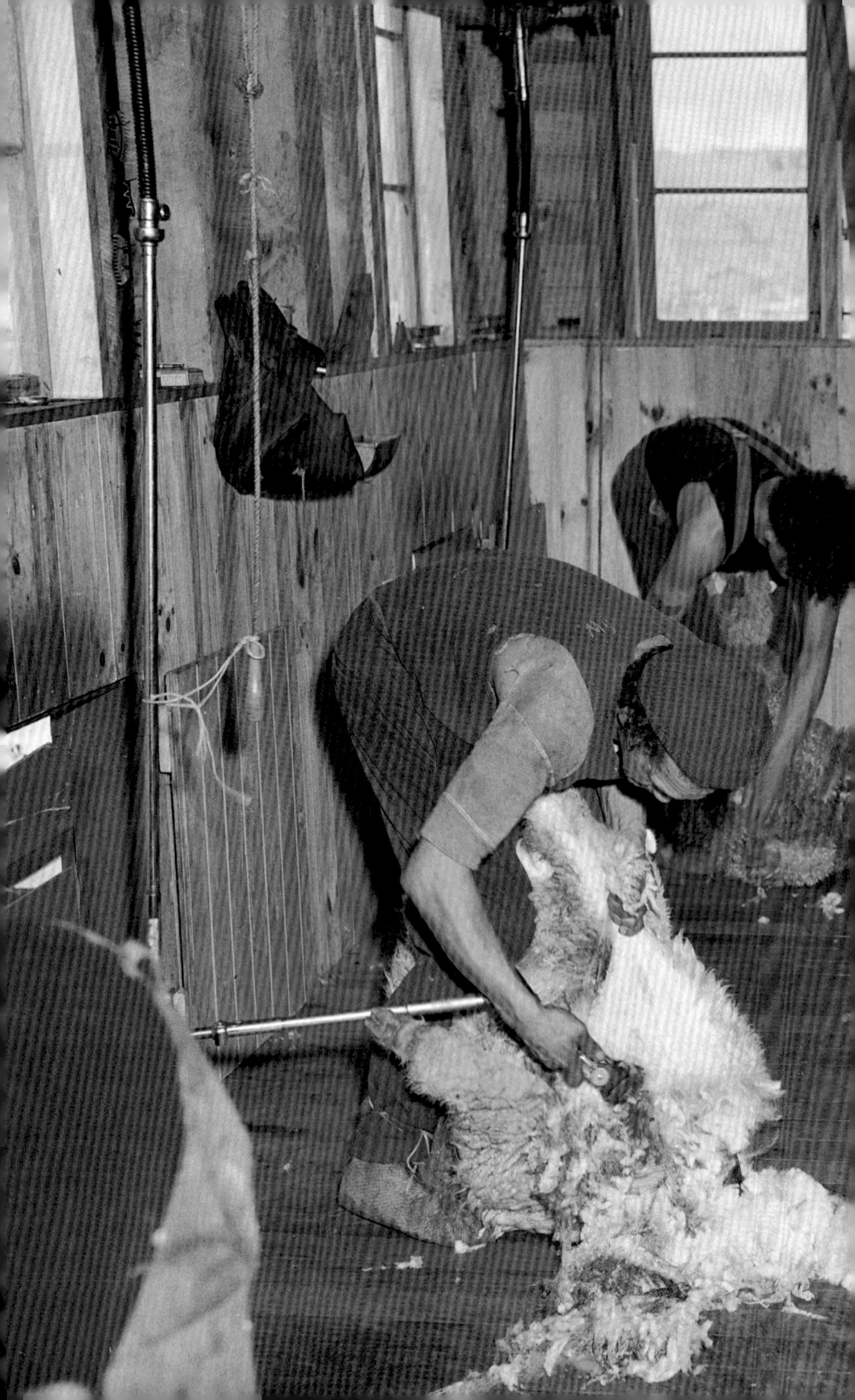

A shed hand waits to remove the fleece from the shearers' stands at Reporoa, between Rotorua and Taupo, in 1953.

While there are several shearing competitions around the country — Gore has had its Southern Shears since 1966 — Te Kuiti, in the Waitomo district, is the self-proclaimed 'Shearing Capital of the World'. As such it is home to the largest shearer in the world, a six-metre high black-singleted roadside statue, unveiled in 1994. A decade on, in March 2004, and to mark the twentieth anniversary of the New Zealand Shearing Championships, the town paid homage to the famous Running of the Bulls in Pamplona, Spain, by initiating its own Running of the Sheep.

Ex-All Blacks Brian Lochore and Colin Meads, wearing Swanndri shirts and gumboots and riding on quadbikes, led a flock of 2000 ewes down the main street. Naturally a permit was required for the event, and the sheep were deprived of their usual feed beforehand to reduce the amount of droppings en route. It proved a popular attraction and the following year's event included the added twist of a competition — to guess the number of sheep involved (correct answer: 2227).

Getting down to more serious business, local shearer David Fagan successfully defended his national title. He had first made his mark on the national scene in 1986 when he won the New

Zealand Golden Shears Open Championship, which he then went on to claim 15 times. Altogether he has won some 550 shearing competitions around the world, including five world titles and 11 United Kingdom Golden Shears Open titles. Most recently, in the 2006 open championships at the Waitomo Cultural and Arts Centre, Fagan came from behind to beat recently crowned Golden Shears champion Dion King of Napier and claim his sixteenth New Zealand championship.

The rules for the Golden Shears World Championships in 2003 required competitors to be 'suitably attired' — that is, no jandals or bare feet. And when it came to the shearing, points would be deducted if any competitor cut a teat and impaired the breeding ability of a ewe, cut or seriously damaged the pizzle of a wether, or cut any sheep such that it required surgical attention.

As well as sections for those operating handpieces, both the Golden Shears and the New Zealand Shearing Champions include competitions for wool handlers. Those individuals also need a quick pair of hands, for their job back in the shed is to get the shorn fleece out of the way of the shearer and on the first stage of its journey to the textile mills and consumers of the world.

Above: **A trio of sheep, with 13 legs between them. Taking**

X

SHEEP FUTURES

The meat and wool of the sheep have long been matters of great economic concern to New Zealand, but the animal's distinctive digestive system has also made the news. As a signatory to the Kyoto Protocol on global warming in 2003, the New Zealand Government revealed plans to research agricultural greenhouse gases, which were believed to be a significant cause of world climate change. That research would be paid for by farmers by means of a levy on the main sources of the problem, their livestock.

The rate for those methane emitters would be about nine cents per sheep and 72 cents per cow and was officially referred to as the flatulence tax, although it was soon dubbed the 'fart tax'. It proved a controversial idea down on the farm, and also in Wellington when an Opposition MP, Shane Ardern, drove a tractor up the steps of Parliament in protest. The Government soon backtracked, announcing that research costs would now be met by an agricultural industry group. The farmers of New Zealand would not be required to cough up for the belching and out-gassing of their animals after all.

Historically, New Zealand has employed shepherds, natural and made-made barriers and dogs to keep its sheep in check. Even so some animals have managed to get away, beyond the reach of the annual muster, and have established feral populations in isolated spots around the country, from northeast Hawke's Bay to the Hokonui Hills in Southland, and on certain islands.

Central Otago has proved a popular spot for renegades, with the recapture in 2004 of a Merino that had been living in mountain caves for six years. Named Shrek, after the green ogre in the animated film, this hermit ram became something of a celebrity and was shorn for an international television audience, his 27.2 kg fleece being auctioned for charity.

Two years later, two more equally hairy rams — half-bred Merino–English Leicesters — turned up at Bendigo Reach in the same region. They had taken advantage of low water levels and wandered out to an island in the Clutha River, where they remained for the next six years. When these lost sheep were found they were also subjected to a dramatic trim.

> ### Woolly words
> FRED DAGG: Black singleted and gumbooted comic farmer character created by John Clarke (born Palmerston North, 1948), who made his first appearance in 1973. Clarke has since become a leading writer and film and television personality in Australia, having left Dagg behind in the heavily subsidised rural sector of New Zealand.

Following the removal of subsidies in the 1980s, New Zealand's sheep population went into free-fall, but in June 2002 it appeared to have levelled off at 39.2 million — a figure last recorded (when it was on the increase) in the mid-1950s. The world sheep population is about 1.03 billion, and overall the animals are outnumbered by humans by some six to one.

By contrast, New Zealanders have been in the minority since the 1840s. The sheep count in 1851 was 233,043, already considerably more than the first human census in 1857–8, which recorded 56,049 Maori and 59,413 Europeans. Currently the New Zealand flock is sixth in the world in terms of size, after China, Australia, India, Iran and Sudan.

The national sheep population may have steadied, but it remains at the mercy of changing markets and technologies. On the positive side, an increase might be stimulated by rising oil prices that make wool a more competitive alternative to synthetic fibres, and draw attention to its natural qualities (water-repellent, good insulator, doesn't burn easily, fibres don't break easily, and, thanks to its natural grease, stays cleaner longer than other fibres) which sheep have long enjoyed.

New Zealand may yet be able to turn other nations on to sheep's other great asset — the taste of lamb and mutton. Any such developments are likely to be lost on the sheep itself. The animal will be too busy to notice, with its head down doing what it's best at: ruminating.

Woolly words

COUNTING SHEEP: A distraction technique, promoted as a cure for insomniacs. The sheer tedium of the task is said to be enough to put anybody to sleep.

Above: The New Zealand farmer — and his sheep — increasingly enjoys the convenience of modern technology, such as the ute. This utility vehicle was allegedly an Australian invention, first appearing there in 1934.

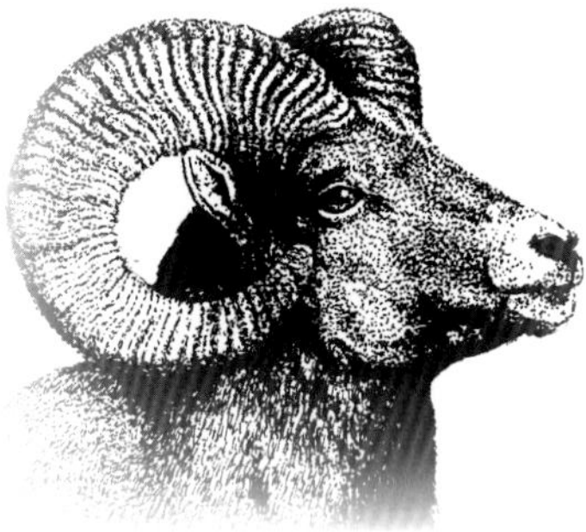

GLOSSARY

Abomasum: The fourth of the sheep's four stomachs.

Bloat: The inability to get rid of excessive gas from the rumen, usually caused by diet.

Blood and bone: Fertiliser; a sheep by-product obtained from the dried and ground residue left after tallow has been extracted in the rendering department at the freezing works.

Board: The area in the woolshed on which the shearers work.

Bot fly (also Head Bot): A fly that deposits tiny larvae on the muzzle or nostrils of sheep. Larvae then migrate to the sinuses and cause irritation, and can result in infection and weight loss.

Bush sickness: A loss of appetite, wasting and anaemia in sheep (mainly) and cattle caused by cobalt deficiency. The lack of this essential trace element is corrected by topdressing with cobalt sulphate, and the use of salt licks for farm stock.

Cast: Condition when a sheep is unable to rise from a lying position on the ground. May be caused by lameness, weakness, lying in a hollow, a heavy or entangled fleece, or advanced pregnancy. The 1937 edition of *Farmcraft* offered one farmer's proven method of getting a cast animal back on its feet. He dug two trenches, for the sheep's front and back feet and just deep enough so the animal could touch the bottom. Within a day or two it would find the use of its legs and be up and away.

Chop (lamb or mutton): So-named because it has been sliced or chopped off. The 1883 *Brett's Colonists' Guide* suggested lamb chops be done on a gridiron, to a 'nice brown', then seasoned with pepper and salt and served with a garnishing of crisped parsley on top of mashed potatoes, vegetables, peas, asparagus, or spinach.

Clip: All the wool shorn from a particular farm, locality or flock of sheep or even, in fact, the whole country.

Crutching: The removal of wool from a sheep's tail and hind legs to prevent fouling. Wool clipped from such parts is known as crutchings.

Dagging: The business of removing matted wool and excrement from the rear quarters of a sheep. When such soiling is still soft or moist, it may be known as green dags.

Dipping: Immersing sheep in solutions of insecticide to protect against ticks, lice and fly-strike has been carried out in New Zealand since the 1850s. There is now concern that many of the chemicals used in the nation's early dips or sheep baths — arsenic-based prior to the 1950s, followed by dieldrin, lindane and DDT, and more recent synthetics — have left toxic residue in the surrounding soil.

Dipping crutch: The 1937 edition of *Farmcraft* suggested a handy device for forcibly submerging sheep could be made from an old file, or rasp, split and shaped by a blacksmith and fitted with a handle.

Docking (also tailing): The removal of a lamb's tail. This process reduces the incidence of fly strike and reduces dags, and is carried out with either a hot iron or a rubber ring. At the same time the lambs may be drenched, ear-marked and vaccinated, and male lambs castrated.

Drafting: The farmer sends the sheep through a race wide enough for only one animal at a time, and operates a drafting gate to separate out the flock. In some cases the farmer may also sort sheep from goats, when some of the latter are rounded up with the muster.

Drenching: The insertion of a drench gun nozzle down the throat of a sheep to administer protection against internal parasites and diseases.

Droving: The driving of large numbers of sheep (or cattle) over long distances.

Dry sheep: A ewe that has not been served by a ram and is not carrying a lamb. Barren ewes are usually disposed of, to maintain a stronger breeding stock.

Eight-tooth (or Full-mouth): A sheep aged about two to four years.

Emasculator: A device for removing the male sheep's testicles, thereby relegating a ram to a wether.

Ewe: Female sheep

Facial eczema: A condition of severe dermatitis caused by a toxic spore found in dead vegetative matter in pastures, especially perennial ryegrass.

Farm: A landholding for rearing animals or growing crops. Less grand than a run or pastoral station.

Flock: A more formal term for a mob or drove of sheep.

Fluke: A grass-borne parasite that establishes itself in sheep's livers.

Fly strike (also called Blowflies): Wool, particularly when dirty, is attractive to flies, resulting in the infestation of the flesh by maggots. Blowflies are also attracted by wounds.

Footrot: An inflammatory bacterial disease, with lameness the most common symptom. Treated by paring and chemicals.

Foot scald: Caused by a soil bacteria that affects the tissue between the sheep's toes. Easier to treat than footrot.

Four-tooth: A sheep aged about 1.5 years to 2 years.

Grass: New Zealand's first sown grass was on land cleared of fern near Kerikeri, Bay of Islands, in 1821. Since then New Zealand sheep have been able to dine out on a wide range of colourfully and curiously named grass types, including meadow foxtail, cocksfoot, timothy, perennial ryegrass, Italian ryegrass, crested dogstail (said to be one of the best sheep grasses in the 1930s), chewing's fescue, brown top, meadow foxtail, white clover and red clover.

Grazier: Common nineteenth-century term for a large-scale sheep farmer.

Gun shearer: An expert at the quick removal of fleeces.

Heading dog: A silent operator which runs to the far side of a mob and brings it back to the master.

Hogget: A lamb becomes a weaner and then, when nearly a year old, a hogget.

Homestead: The headquarters, including the owner's residence and ancillary buildings, on a station.

Huntaway dog: One that enjoys barking as it drives sheep away from its master.

Hydatids: The sheep is the most common intermediate host in the lifecycle of this tapeworm which lives in the gut of a dog. The parasite probably came to New Zealand with sheep in the nineteenth century.

Keds (or ticks): Wingless biting flies with sucking mouthparts which feed on blood and cause sheep to go off their food, lose condition and grow less wool. Keds also stain wool with their excrement and thus reduce its value.

Kemps: The course, rough hair in wool, common on more primitive (less domesticated) breeds of sheep.

Killer: Not a vicious sheep but one identified for killing and eating on the farm. Also known as a mutton.

Lamb (animal): A newborn sheep until it is weaned, at about the age of five months. When it no longer drinks from its mother it becomes a weaner. New Zealand lambs are born in late winter and spring (August–October).

Lamb (meat): The 1883 *Brett's Colonists' Guide* recommended that lamb is cooked fresh and roasted 'at a brisk fire'. A leg would take about an hour and a half by that method, while a forequarter and a loin and saddle would require about two hours. The *Guide* offered recipes for boiled lamb, lamb chops and lamb cutlets. Nearly a century later, a 1979 New Zealand recipe book recommended a leg of lamb be baked in a moderate oven (180°C) for 2–2½ hours, with potatoes, butternut or kumara added to the dish later and to be served with mint sauce or, alternatively, mushroom gravy.

Lanolin: The pale yellow natural oil found on sheep's wool. As well as waterproofing the sheep, it has anti-fungal and antibacterial properties which protect the animal's skin from infection. Used in the pharmaceutical and cosmetic industry, lanolin prevents skin from drying and cracking, and is the reason sheep shearers have such soft hands.

Lice: Permanent ecto-parasites which spend their lives on a sheep, and are usually spread by contact between animals. Three types infest New Zealand sheep: one biting and two sucking species.

Maiden ewe: A young ewe having its first lamb.

Matagouri: A thorny native bush or small tree, often prickly, especially widespread in the South Island.

Mob: An unruly assembly, usually of sheep or cattle. Less formal than a flock.

Mouflon: A wild sheep (*Ovis musimon*) inhabiting the mountains of Sardinia, Corsica, etc. Thought to be the ancestor of the domesticated sheep.

Mountain mutton: New Zealand poached venison, when deer were protected and could only be shot in season.

Mountain oysters: Lambs' testicles fried in breadcrumbs are said to be a delicacy, resembling brains or whitebait.

Mustering: The muster is the rounding up of sheep (and sometimes a few goats) from around the farm and bringing them back to the yards and holding paddocks. Carried out prior to weaning or shearing.

Mutton: The flesh of the sheep, and also a sheep to be killed on a farm to provide meat. In her classic *All About Cookery* (1903), Mrs Beeton identified the five joints from a side of mutton: the leg, loin (together forming the saddle), the shoulder, neck and the breast. The leg and much of the loin constitute the haunch of mutton, then a popular dish at public dinners and special events. Mrs Beeton offered instructions for the carving of the main joints, along with recipes for baked minced mutton, boiled breast of mutton, broiled mutton, mutton broth, mutton chops, mutton collops, mutton croquettes, curried mutton, mutton cutlets (mashed or Italian), braised fillet of mutton, haricot mutton, hashed mutton, ragout of cold neck mutton, mutton pudding, mutton pie and mutton soup.

A variation on mutton is the New Zealand colonial goose. A leg or shoulder of boned mutton roasted with a savoury stuffing was so named for its resemblance to a stuffed goose. Described as a true national dish and dating from the nineteenth century, a 1979 recipe recommended this 'goose' be served on a platter surrounded by baked vegetables and with gravy, mint sauce or tart jelly.

Omasum: The third of the sheep's four stomachs.

Paddock: A fenced piece of land, of any size.

Pastoralism: The practice of large-scale farming, as carried out by a pastoralist.

Pizzle rot: A bacterial infection of the ram's sheath. Ammonia in the urine then causes irritation and ulceration, resulting in the ram losing both the desire and ability to breed. Pizzle was an early term for the penis, and used by Shakespeare in *Henry IV* when Falstaff referred to a 'bull's pizzle'.

Pulpy kidney: A bacteria produces a toxin which causes death, mainly in single young lambs. Ewes can be vaccinated prior to lambing.

Quadbike: Farmers once got around their sheep on horseback, but horses have been mostly replaced by farmbikes and quadbikes. There are now about 70,000 four-wheel ATVs (all-terrain vehicles) on the farms of New Zealand.

Ram harness: A device which indicates whether a ewe has been serviced by a ram, incorporating a crayon which leaves a mark on the recipient. Crayons are available in a range of colours, and the mark is fully removable and in accordance with the Animal Act (1967).

Rams: Male sheep, with all parts intact.

Reticulum: The second of the sheep's four stomachs.

Ringer: The shearer who manages the highest tally for the day; the fastest in the gang.

Rouseabout (also Rousabout, Rousie): General hand in the shearing-shed.

Rumen: The first and largest of the sheep's four stomachs.

Run: Sheep or cattle station, or open grazing land, as owned, managed or leased by a runholder. In 1851 Frederick Weld referred to a 'run' or a 'sheep-walk'.

Scab: A contagious disease caused by mites feeding on the surface layers of the sheep's skin. Treated by dipping.

Scrapie: This sheep disease was imported and appeared in a South Island flock in 1952, and was later eradicated.

Shearing: Traditionally takes place in early summer (November–December).

Sheep measles: Similar to hydatids; the sheep plays an intermediate role in the lifecycle of this tapeworm.

Sheep pellets: Pelletised sheep manure and wool waste for fertilising and aerating the soil.

Six-tooth: A sheep aged about 2–3 years.

Slink: The skin of a dead lamb. Natural casualties during lambing season, caused by harsh weather conditions, are collected and their skin processed into sought-after leather.

Slipe: Wool taken from the sheep's back after slaughter.

Squatter: A (usually wealthy) pastoral farmer distinguished by the large scale of his holding.

Also a member of a social and political elite.

Squattocracy: Party of wealthy sheep farmers who were a significant political force in New Zealand in the latter half of the nineteenth century.

Stand: The position occupied by a shearer on the board in the woolshed.

Staggers, ryegrass: Usually occurs when rain follows a prolonged dry spell. Toxin from a fungi growing in the seed heads and lower stems of ryegrass causes sheep to experience head wobbles, stiff legs and what appear to be seizures. Most sheep recover without the farmer needing to call the vet.

Staple: The length of the longer fibres in a 'staple' of wool in its natural condition. Staple length is one of the five important means of measuring and determining the value of a wool sample, the others being its fibre diameter, yield, staple strength and colour.

Station: A large grazing property with buildings and stock.

Stock (and station) agent: An individual, employed

by a company, who advises on farming matters and arranges the buying and selling of livestock. Some of the big names in the business around the country in the mid-1970s were Dalgety, Newton King, Williams & Kettle, Wrightson NMA and Pyne, Gould, Guinness.

Straggler: A sheep that misses — or eludes — the main muster and is rounded up later.

Swordgrass: New Zealand plants with sword-shaped leaves, such as the toetoe (commonly misspelt toitoi).

Tripe: A dish made from the large stomach of ruminating animals. As Mrs E. Barrington pointed out in her 1940 *Centennial Recipe Book*, 'Many people dislike tripe on principle'. She assured readers there were 'several interesting things one can do with tripe', and offered recipes for creamed tripe ('Sounds much nicer than the usual idea'), tripe cutlets and stuffed tripe roll. Even Shakepeare had something to say about the dish, in *The Taming of the Shrew*: 'How say you to a fat tripe finely broiled?' A modern variation, said to be unsuitable for the squeamish, is the Lancashire delicacy of tripe and onions.

Tupping: In the nineteenth century 'tup' meant to butt

like a ram or, more engagingly, to 'cover', as the male sheep does during mating. Even Shakespeare referred to this essential pastoral activity, in *Othello*:

> *'Even now, now, very now,*
> *an old black ram is tupping your white ewe.'*

Nearly 360 years later, in 1980, in his landmark play *Foreskin's Lament* New Zealander Greg McGee gave the rugby coach character the name Tupper. Tupping takes place in New Zealand paddocks in autumn (March–May).

Tussock: Prior to European settlement, most of the eastern side of the South Island was covered with areas of tussock which contained many palatable (to sheep) grasses and herbs, and provided an instant pasture. There were four types of tussock: red, snow, silver and fescue.

Tutu: A native shrub or small tree found throughout New Zealand and the Chatham Islands. Poison in the sap and seeds resulted in a large number of stock deaths in the early days of European settlement. It is also dangerous to humans, and referred to as 'toot'.

Two-tooth: A sheep aged about 12 months to 18 months.

Weaner: A young sheep that no longer drinks from its mother. When it is nearly a year old it becomes a hogget.

Wether: A castrated ram.

Wool: Strictly speaking, the soft wavy or curly hypertrophied undercoat of various hairy mammals, made up of a matrix of cylindrical keratin fibres covered with minute overlapping scales.

Wool grower: Unfairly perhaps, a wool grower is not a sheep but a person who raises sheep for wool.

Wool press: A device, previously hand-powered with ratchets and pulleys but now electrical, for compressing the newly sheared wool into bales for transportation from the farm to the wool stores.

Works, the: The freezing works, where most New Zealand sheep are destined after their days on the farm.

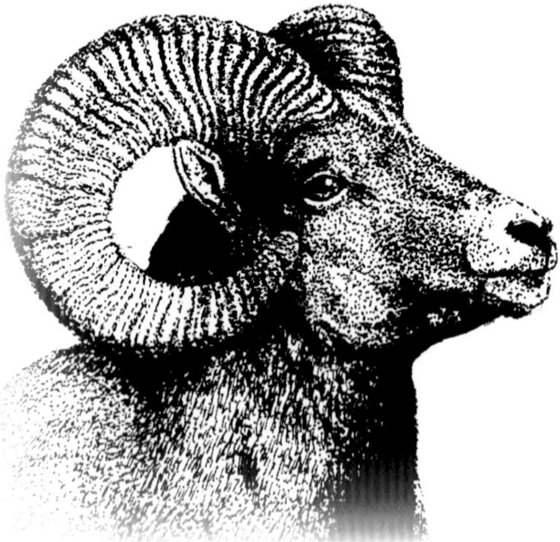

BIBLIOGRAPHY

Barrington, Mrs E., *Centennial Recipe Book and Cooking by Electricity*, Taranaki Daily News Co Ltd, New Plymouth, 1940

Beeton, Mrs (Isabella Mary), *All About Cookery: a collection of practical recipes, arranged in alphabetical order*, Ward, Lock, London, 1903

Begg, A. Charles and Neil C. Begg, *James Cook and New Zealand*, Government Printer, Wellington, 1970

Caras, Roger A., *A Perfect Harmony: the intertwining lives of animals and humans throughout history*, Simon & Schuster, New York, c.1996

Clutton-Brock, Juliet, *A Natural History of Domesticated Mammals*, Cambridge University Press, Cambridge, 1987

Cumberland, Kenneth B., *Landmarks*, Reader's Digest Services Pty Ltd, New South Wales, 1981

Duff, Oliver, *New Zealand Now*, George Allen & Unwin Ltd, London, 1956

Goddard, W.H., *Best New Zealand Dishes*, Price Milburn and Company, Wellington, 1979

Gustafson, Barry, *His Way: A Biography of Robert Muldoon*, Auckland University Press, Auckland, 2001

Hereford, P.S.E., *The New Zealand Frozen Meat Trade*, New Zealand Publishing Co Ltd, Wellington, 1932

Kalaugher, J. P., *Gleanings from Early New Zealand History*, Unity Press, Auckland, 1950

Keith, Hamish, *All Our Yesterdays*, Reader's Digest Services Ltd, New South Wales, 1984

Leys, Thomson W. (editor), *Brett's Colonists' Guide*, H. Brett, Auckland, 1883

Loach, Cyril, *A History of the New Zealand Refrigerating Company*, Caxton Press, Christchurch, 1969

McLauchlan, Gordon, *The Farming of New Zealand*, Australia and New Zealand Book Company, Auckland, 1981

McLean, Gavin, in *Frontier of Dreams* (edited by Bronwyn Dalley and Gavin McLean), Hodder Moa Beckett, Auckland, 2005

McLintock, A.H. (editor), *An Encyclopaedia of New Zealand*, Government Printer, Wellington, 1966

Nobbs, K., *The First Introduction of Sheep into New Zealand* (unpublished, Auckland Central Library)

Ogonowska-Coates, Halina, *Boards, Blades & Barebellies*, Benton Ross, Auckland, 1987

Orsman, H.W. (editor), *The Dictionary of New Zealand English*, Oxford University Press New Zealand, Auckland, 1997

Palenski, Ron, 'Bowen, Walter Godfrey 1922–1994'. *Dictionary of New Zealand Biography*, updated 7 April 2006 http:www.dnzb.govt.nz/dnzb

Parkes, John, Advances in New Zealand mammalogy 1990-2000: feral livestock, *Journal of the Royal Society of New Zealand*, Vol. 31, No. 1, March 2001.

Parry, Gordon, *N.M.A. The Story of the First 100 Years*, N.M.A., Dunedin, 1964

Ponting, Keith, *Sheep of the World*, Blandford Press Ltd, Dorset, 1980

Ryder, M.L., *Sheep & Man*, Gerald Duckworth & Co. Ltd, London, 1983

Simpson, Tony, *A Distant Feast: The Origins of New Zealand Cuisine*, Godwit, Auckland, 1999

Smith, Henry B., *The Sheep and Wool Industry of Australia and New Zealand*, Whitcombe and Tombs, Auckland, 1927

Sutherland, Allan, *New Zealand Famous Firsts and Related Records*, Allan Sutherland, Auckland, 1961

Thompson, A.S., *The Story of New Zealand*, J. Murray, London, 1859

Tonson, A. E., 'The Arms of the Rt Hon Sir Keith Jacka Holyoake, KG, GCMG, CH, LLD', *The New Zealand Armorist*, No. 21, December 1982, 4–6.

Weber, Rex A., *Wool Man*, Cape Catley, 1992

Williams, Des, *Top Class Wool Cutters*, Shearing Heritage Publications, Hamilton, 1996

Code of Recommendations and Minimum Standards for the Welfare of Sheep, Animal Welfare Advisory Committee, Ministry of Agriculture, Wellington, 1996

Farmcraft, Vol. II, Wilson & Horton Ltd, Auckland, 1937

Farming in New Zealand, Department of Agriculture, Vol 1, Wellington, 1950

Manufacturing in New Zealand, Cranwell Publishing, Auckland, 1959

New Zealand Agriculture: A Story of the Past 150 years, NZ Rural Press Ltd and Hodder & Stoughton, Auckland, 1990

New Zealand Official Year Book, Government Printer, Wellington (various editions)

New Zealand Sheep and their Wool, Wools of New

Zealand, Wellington, 1994

Official Handbook of the Auckland Industrial and Mining Exhibition 1898–99, Geddis and Blomfield, Auckland, 1898

Official Record of the New Zealand Industrial Exhibition, 1885, Government Printer, Wellington, 1886

The New Zealand Handbook, or Guide to the Britain of the South, E. Stanford, London, 1879

The Journal of Agriculture, New Zealand Department of Agriculture, Industries and Commerce, Wellington, Vol. xiv, No. 1, 20 January 1917

The Settler's Handbook of New Zealand, Government Printer, Wellington, 1902

Image Credits

Sheep breed images courtesy of Meat & Wool New Zealand and the New Zealand Sheepbreeders Association

Other images courtesy of:

Alexander Turnbull Library: 10 (Mrs Fox Collection), 30, 39 (The Press [Christchurch] Collection), 42, 44, 48, 52, 54, 58 (Making New Zealand Collection), 63 (New Zealand Railways Collection), 67, 68, 70 (B Davis Collection), 74, 80 (Steffano Webb Collection), 84 (Howard Collection), 87 (James McAllister Collection), 96, 102, 106 (Steffano Webb Collection), 109, 112 (J Atkinson Collection), 117 (Steffano Webb Collection), 119, 121, 122 (Adkin Collection), 125 (SC Smith Collection), 127 (James McAllister Collection), 130 (John Pascoe Collection), 135 (FA Hargreaves Collection), 136, 138, 142 (John Pascoe Collection), 146 (James McAllister Collection), 149, 152 (New Zealand Free Lance Collection), 155 (John Pascoe Collection), 156 (SC Smith Collection), 159 (Gordon Burt Collection), 166 (Evening Post Collection), 168 (Morrie Hill Collection), 170 (Northwood Collection), 173, 174, 177, 180 (New Zealand Free Lance Collection), 184 (Archives New Zealand: National Publicity Studios Collection), 188

Library of Congress: 12, 21, 33, 36

Natural Science Image Library: 163 (Photograph: GR 'Dick' Roberts), 193 (Photograph: Peter E Smith)

US Fish and Wildlife Service National Conservation Training Centre: 15 (Photographer: Gary Kramer)

US Department of Agriculture Agricultural Research Service: 17 (Photograph: Scott Bauer), 29

Washington Department of Fish and Wildlife: 24

INDEX

Aberdeen Angus 145
acclimatisation 32,
Acland, John 51,
  53, 57, 121
Adam and Eve 20
Addington 142
Adelaide Bell 56
Aesop 16
Agrodome, The 182
Albion Shipping
  Company 100
Antarctica 32
antelopes 26
apparel manufacture
  83
Arapawa Island 35, 36
Ardern, Shane 190,
Argentina 115
Aries 22
arsenic 53
Artiodactyla 26
Ashburton 104, 117
assistance
  schemes 165
Auckland 148
Auckland Agricultural
  and Horticultural
  Society 62
Auckland Farmers
  Freezing
  Company 106
auction 154
Australia 36, 37, 39,
  41, 46, 51, 53, 54,
  60, 73, 79, 83, 86,
  89, 115, 191, 192
Avon River 51, 52

Baring Head 49
Bay of Islands 30, 32,
  40, 42, 46, 171
Belfast Freezing
  Works 105, 117
Belgium 169
bellwether 167
Bendigo Reach 191
Berkshire 91

Bible, the 20, 22
Bidwill, Charles 47, 49
bighorn mountain
  sheep 24, 27
Blackhead Persian 92
blackleg 137
Bligh, William 38
blowflies 137
Bluff 112
Border Leicester 76,
  78, 82, 83, 86,
  87, 93, 94, 116
Borderdale 28, 83, 86
Botany Bay 36
Bounty 38
Bovidae 26
Bowen, Godfrey
  178–183
Bowen, Ivan 179,
  182, 183
branding 124
breeding 18
Brees, Samuel
  Charles 44
Brydone, Thomas
  100, 101
bullswool 176
bully beef 103
Burnside 105
butcher 101
Butler, Samuel 57, 58
butter 36, 66, 99, 101
Byron, Lord 153

Caduceus 99
Cain and Abel 20
camel 26
camouflage 28
Campbell Island 77
candle making 98, 110
Canterbury 44, 48,
  50–53 56, 57, 61,
  82, 83, 86, 99, 100,
  104, 109, 115–118,
  132, 148, 151, 165
Canterbury Flax
  Spinning Weaving
  and Fibre
  Company 66
Canterbury Frozen
  Meat and Dairy

Produce Co
  Ltd 105, 117
Canterbury Meat
  Company 96, 98
Cape Campbell 50
Cape Horn 105
Cape of Good Hope
  32, 34, 37
Cape sheep 36, 75
Capricorn 22
Caras, Roger 16
carbolic acid 53
carpet manufacture
  78, 83, 86,
  90, 94, 169
castration 131, 137
cattle 26, 34, 41,
  56, 151, 167
Chatham Islands 77
cheese 66, 92, 104
Cheviot 78, 79, 83,
  89, 90, 118, 183
Cheviot Estate
  118, 119
Cheviot Hills 89, 119
China 169, 192
Christchurch 117,
  142, 144
Clarke, John 191
Clifford, Charles
  44, 47, 49
Cloudy Bay 50
Clutha River 191
coat of arms 143–145
Collies 61, 132
Columbus,
  Christopher 31
Cook, James
  32–36, 75
Coop, Ian 78
Cooper handpieces
  176
Coopworth 28, 78,
  86, 94, 145, 166
Corriedale 28, 74–77,
  79, 83, 86, 95,
  139, 150, 166
Corriedale Station 79
Cotswold 95
Cruise, Richard 40
cud 27

Cyclopaedia of Useful Knowledge 69

Dagg, Fred 191
dags 89, 137, 141
dairy products 162
Dance, Nathaniel 33
Daroux Emasculator 131
Dartmoor 95
Darwin, Charles 60
Davidson, William Saltau 100, 101
Deans, John 51, 52
Deans, William 51, 52
deregulation 168
Destruction of Sennachrib, The 153
dipping 122, 126, 128, 133
Discharged Soldiers Settlement Act 1915 148
Discovery 34
diversification 168
docking 131, 137
dogs 14, 15, 17, 18, 49, 55, 130, 132, 190
dog-trialling 132,
domestication 14, 17–20, 23, 28, 34, 167
Donald & Son 67, 68
Dorper 92
Dorset Down 88, 91, 94
Dorset Horn 86, 92
Douglas, Roger 164, 165
drafting 163
drenching 129, 137, 160
Dromedary 40
drought 53
Dry, Sir Francis 82
Drysdale 28, 82
ducks 34, 101, 104
Duff, Oliver 151, 153
Duke of Edinburgh 168

Dunedin (city) 69, 70, 105, 145
Dunedin (ship) 100–105, 108–110, 114, 164
Dusky Bay 32

earmarking 124, 137
East Friesian 92
eggs 101
Endeavour 32, 34
England 28, 34, 38, 86, 89, 91, 95
English Leicester 76, 79, 82, 88, 94, 191
Erewhon 58, 60
European Economic Community 160, 161
ewes 19, 32, 83, 87, 90–92, 94, 100, 115, 150, 158, 165, 183, 187
exporting 46, 63, 64, 79, 89, 96–122, 133, 139, 158, 160, 161, 169

facial eczema 91, 140
Fagan, David 186, 187
Fairfield 117
Fairlie 61
fart tax 190
fertiliser 80, 131, 148, 162
Finland 90
Finn 90
Finnish Landrace 90
Finnsheep 90
Flaxbourne 50
flystrike 140
foot and mouth 114
footrot 53, 83, 137
footscald 140
Forbes, George 118
France 28, 169
freemarketeers 162
freezing works 99, 102, 105, 108–110, 112, 117, 155

Friday (dog) 61

gadfly 55
Gear Meat Preserving and Freezing Company 105, 108, 168
geese 34, 101
George III 37–39
Germany 65, 92, 94, 169
Gisborne 109, 178
goat 16–19, 21–24, 34, 163, 167
Golden Shears 181, 187
Gore 186
Gotland Pelt 93
greenhouse gases 190
Grey, Lady 62
Grey, Sir George 62
Grigg, John 104, 105, 117, 148
gumboots 186

Hall, William 40
Hamilton 167
Hampshire 91, 93
Hampshire Down 76, 91
Harford, George 182
Hastings 109, 144, 178
Hawke's Bay 88, 151, 181, 183, 190
heading dog 132
HL Rutgers 56
Hokonui Hills 190
Holland 89, 92
Holyoake, Sir Keith 79, 148
Home Leicester 116
Hong Kong 169
horses 34, 41, 132, 166
Huiarua 135
huntaway 132, 133, 182

Illustrated New Zealand News 173
India 37, 192
industrialisation 31
Invercargill 112, 145
Iran 20, 192
Iraq 19, 20
Isle of Wight 103
Israel 22
Italy 28
Ivey Hall 80
Ivey, WF 80

Japan 169, 182
Jerusalem 21
Jubilee 108

Kaiapoi Woollen Manufacturing Company 69
Kenya 139
Kerikeri 40
Kerry Hill 95
King, Dion 187
King, John 40, 46, 171
kiwi 148
knitting 86, 88
Korea 154, 169
Krushchev, Nikita 182
Kyoto Protocol 189

Lake Tekapo 61
Lake Wairarapa 49
lamb (meat) 46, 61, 86, 89, 91, 94, 99, 101, 104, 106, 110, 114, 115, 117, 160, 169, 192
lambs 76, 87, 88, 94, 111, 137, 150
land clearance 18, 62, 90
leasehold land 47
Leicester 62, 82
Levels Station 60, 100
Levin 122
lice 55, 124, 125, 131
Lincoln 76, 79, 82, 83, 88, 90, 100

Lincoln University 78, 80, 120, 179
Lister handpieces 176, 177
Little, James 79
liver fluke 55
Livestock Incentive Scheme 162
Lochore, Brian 186
London 103, 104, 106
Longbeach 117, 148
Longbeach Station 104
Luxembourg 169
Lyttelton 51, 60, 118, 139

Macbeth 35
Mackenzie Country 60
Mackenzie, James 60, 61, 100
maggots 55
Makarewa 112
Mana Island 45, 46
Manawatu 179
Mangere Farmers' Club 148
manure 98
Marlborough 36, 44, 50, 62, 83
Marsden, Samuel 38–40, 42
Massey University 79, 82, 84, 179
Massey, William 148
Masterton 67, 181
matagouri 55
Mataura 104
McArthur, John 37
McLauchlan, Gordon 57
McLean, Gavin 105
Meads, Colin 186
meat 8, 18, 19, 47, 66, 86, 87, 91–93, 97–99, 161, 162, 167, 189,
Melbourne 100
Mesopotamia station 57–60

milk 19, 35, 92, 101, 167
Ministry of Agriculture and Fisheries 134
missionaries 30, 40, 41
moa 144
mohair 167
Mosgiel Woollen Mill 70
mosquitoes 49
motorbikes 132, 166
Mouflon 27
Mount Peel 53, 121
Muldoon, Robert 162
mustering 90, 135, 190
mutton 35, 46, 61, 86, 101, 103–106, 108, 110, 111, 114, 115, 117, 160, 169, 192
muttonbird 120
myxomatosis 64

Napier 144, 187
Nauru 131
Nelson 47, 51, 53, 60
Nelson, William 109
New Plymouth 131
New South Wales 36–39, 56, 65, 144, 171
New Zealand Agricultural and Pastoral Societies (A&P) 179, 181
New Zealand and Australian Land Company 90, 100
New Zealand Company 46
New Zealand Halfbred 28, 77, 82, 95, 150
New Zealand Industrial Exhibition 108
New Zealand Now 151, 153
New Zealand Refrigeration Company 105
New Zealand Shearing Championships 186

New Zealand Wool Board 179
New Zealand Wool Commission 154
Ngaio 44
Northam 99
Northland 183

Oamaru 101
Ocean Beach 109, 112
Ocean Island 131
Okaihau 183
Opiki 179
Orari Gorge Station 53
Organisation for Economic Co-operation and Development (OECD) 104
Otago 51, 57, 62, 70, 79, 83, 151, 190
Ovis aries 27
Ovis aries 34
Ovis canadensis 27
Ovis orientalis 27
oxen 13
Oxford Down 92

Palmerston North 84, 191
Pareora 117
Parramatta 38, 41
pastoralists 47, 49, 56, 116
Pecora 26
penguins 104
Peren, Sir Geoffrey 84
Perendale 28, 78, 79, 84, 90, 94, 94, 166
Perendale, Sir Geoffrey 79
Petone 46
Petre, Charles 47
pheasant 101
phosphate 80, 120, 131
pigs 26, 45, 49, 72, 104
Pitt Island 77

pizzle rot 140
Plymouth 35
Police Offences Act 1884 124
Poll Dorset 86
Polwarth 83, 95
Porirua 44, 45
Port Chalmers 101, 102, 104
Port Nicholson 51
Port Underwood 50
preserved meat 98, 99, 111
pukeko 104
Puketitiri 183
pulpy kidney 137

Queen Charlotte Sound 32, 35

rabbits 62–65, 101, 104, 132, 168
rabbit boards 64
rabbit calicivirus 64
Rabbit Nuisance Act 1876 62
raddle 124
Rakaia River 57
rams 32, 83, 94, 100, 133, 144, 150, 190
Rangihoua 40
Rangitata River 53, 57
rationing 155
refrigeration 8, 63, 76, 86, 93, 96, 99–103, 105, 106, 108, 110, 114, 115, 139
Reporoa 185
Resolution 32, 34
reticulum 27
Riccarton 52
Rimutaka Range 49
Robinson, William 119
Rogernomics 165
Romney 28, 77–79, 82, 83, 86, 88, 90, 94, 150, 158, 166, 183
Romney Marsh 76
Roscommon 95

Roslyn Mills 69
Ross and Glendining 69
Rotorua 182, 185
ruminants 26, 151, 192
runholder 54, 55, 65, 98
Running of the Sheep 186
ryegrass 140
Ryeland 93

saleyards 142, 147
salt-lick 128
saltpetre 96
Samwell, David 34
Saudi Arabia 133, 169
scab 53, 125, 131
Scotch Blackface 95
Scotland 28
Second World War 131, 150, 154, 168, 178, 181
Shakespeare, William 35
Shaw, Savill & Albion 103
shearing 40, 54, 65, 97, 124, 132, 170–187, 191
  handpieces 176, 187
  shed 182
sheep, coloured 94, 95
sheep's tongue 104, 108
sheepskins 111, 157
shepherds 12, 21, 35, 72, 132, 134, 190
shepherd's pie 23
Ship Cove 34
Shrek, the sheep 190
Shropshire 76, 93
slipes 94
soap making 98, 110
Sockburn 109
South Africa 36, 92
South Dorset Down 94

South Hampshire 28, 93
South Suffolk 28, 86
Southdown 76, 86, 88, 91, 93, 94
Southern Alps 82
Southern Frozen Meat and Produce Export Company 112
Southern Shears 186
Southland 60, 62, 109, 151, 190
Soviet Union 28, 182
Spain 37, 38, 65
speargrass 55
Speed 56
squatters 50, 72
staggers 140
stations 57
Stewart Island 51
Stock Act 1893 124
stock auction 46
Strathleven 99
subsidies 161, 162, 164, 165, 191
Sudan 192
Suffolk 86, 88, 91
Supplementary Minimum Prices 162
Sussex 91
Swanndri 186
Sweden 93
swedes 126, 136
Sydney 38, 40, 45, 47, 50

tallow 111, 124
Tamaki 148
tanning 110
Taranaki 147
Tararua Range 49
Taupo 185
Te Kuiti 186
Te Wahapu 46
tetanus 137
Texel 89
textile manufacture 38, 66, 69, 82, 87, 90, 93, 187

Thala Dan 139
Thibenzole 160
ticks 124, 125, 131
Tigris River 57
Timaru 117
Tokomaru Bay 135, 178
Tomoana Freezing Works 109
topdressing 120, 131
Totara Estate 101
Treaty of Rome 160
tripe 27
Tripp, Charles 51, 53, 57
Tripp, George 121
Tunis 95
tupping 133
turkey 34, 101,
tussock 50, 53, 55, 57, 62
tutu (plant) 32, 55
Two Gentlemen of Verona 35

United Kingdom 31, 65, 66, 114, 115, 169
utility vehicle (ute) 193

Vavasour, William 44, 47, 49
venison 108

Waikato 120
Waimakariri River 48
Waimate 131
Waimate North 30
Wainuiomata 130
Wairarapa 44, 47–50, 83, 88, 151
Wairarapa Young Farmers' Club 181
Wairoa 129
Waitangi 40
Waitomo Cultural and Arts Centre 187
Wales 28, 95, 183
Wanganui 145
Ward 50

Ward, Joseph 112
waterfront dispute 1951 154
Waterhouse, Henry 37, 38
Weekly News 128, 129, 172, 176
Weld, Frederick 48, 49
Wellington 45, 46, 49, 51, 105, 108, 116, 144, 151, 157
Wensleydale 95
West Indies 34
Western Star 56
Westfield Freezing Works 155
wethers 140
whaling 45, 46
White Headed Marsh 94
Wiltshire 89, 91
wolf 15, 22
Wolseley handpieces 176, 179
wool 8, 19, 31, 38, 47, 50, 53, 57, 62, 65, 66, 72, 76, 79, 82, 83, 86–89, 92, 94, 97, 98, 111, 153, 154, 158–162, 164, 165, 169, 189
 bales 66, 128, 159, 164, 172, 178
 brokers 178
 classing 178
 handling 187
 press 67, 68
 scour 110
 sheds 172, 181
Wool Proceeds Retention Fund 154
woollen mills 31, 69, 70, 110
Woolworths 183
worms, intestinal 160
Wright, John Bell 45, 46

Yorkshire 38

zodiac 22

IN THE SHEARING SHED